THE PHYSICS OF HUMAN MOVEMENT

BY

R. L. Page, Ph.D., B.Sc., M.Inst.P.

Director of the Science and Technology Education Centre,
School of Education, University of Bath

A Division of Pergamon Press

A. Wheaton & Company Limited
A Division of Pergamon Press
Hennock Road, Exeter EX2 8RP

Pergamon Press Ltd
Headington Hill Hall, Oxford OX3 0BW

Pergamon Press Inc.
Maxwell House, Fairview Park, Elmsford, New York 10523

Pergamon of Canada Ltd
75 The East Mall, Toronto, Ontario M8Z 2L9

Pergamon Press (Australia) Pty Ltd
19a Boundary Street, Rushcutters Bay, N.S.W. 2011

Pergamon Press GmbH
6242 Kronberg/Taunus, Pferdstrasse 1, Frankfurt-am-Main, West Germany

---

Selected Topics in Physics

Editor: Brian E. Woolnough

Exciting Electrical Machines	E. R. Laithwaite
What Is Energy?	A. W. Wilson
Light and Life	A. McB. Collieu
Waves and Oscillations	P. G. Parkhouse
The Physics of Human Movement	R. L. Page

Copyright © 1978 Ray Page

All rights reserved. No part of this publication may be reproduced, stored in a retrieval system, or transmitted, in any form or by any means, electronic, electrostatic, magnetic tape, mechanical, photo-copying, recording or otherwise, without permission in writing from the publishers.

First edition 1978

Printed in Great Britain by A. Wheaton & Co. Ltd, Exeter

ISBN 0 08 021362 6

Contents

	Acknowledgements	iv
	Preface	v
Chapter 1	Velocity analysis of human movement	1
Chapter 2	Human forces	13
Chapter 3	The centre of gravity of the human body	31
Chapter 4	Conserving momentum	45
Chapter 5	The physics of ball games	57
	Suggestions for projects and investigations	69
	References	71
	Index	73

Acknowledgements

Some of the material in this book has appeared in the *School Science Review* and the *Physical Education Journal* and is used here by permission of the Association for Science Education and the Physical Education Association. J. Smith is thanked for permission to quote results from his project; A. Philpott, St Paul's College of Education, Cheltenham for permission to quote results gained by him using a trifilar cradle; C. B. Daish for permission to quote from his study on the effectiveness of the golf club head in Chapter 5 and R. C. Haines, Technical Development Department, International Sports Co. Ltd, Barnsley for the data supplied on the changes in coefficients of restitution for tennis balls.

The following are acknowledged for providing photographs or material for diagrams:

Prof. H. E. Edgerton, M.I.T. (Fig. 1.9)
BBC (Fig. 1.10)
Natural Film Archives (Figs. 1.13 and 1.14)
Prof. T. K. Cureton, late of Springfield College, Mass., U.S.A. (Table 4 and Fig. 2.14)
H. Payne, Dept of Physical Education, University of Birmingham (Figs. 2.17–2.22 and References 7, 9 and 10)
Lloyds Instruments Ltd, Banbury and Royal Aircraft Establishment, Farnborough (Fig. 2.23 and permission to describe their Stanmore Sandals)
G. Dyson (Figs. 2.2, 3.2, 3.3, 3.7, 4.9–4.12 and 4.18)
J. Williams, The John of Gaunt School, Trowbridge (Figs. 3.9 and 3.11)
Editor, *Research in Physical Education* (Figs. 3.15–3.18)
K. Lane, Dept of Physical Education, University of Edinburgh (Fig. 3.19)
R. N. Harris, Springfield College, Mass., U.S.A. (Fig. 3.20)
G. A. Carter, School of Education, University of Bath (Fig. 5.10)

Preface

The mechanics section of many physics courses frequently appears less interesting than other parts of the course. One reason for this is undoubtedly that the relevance of this area of physics to modern transport technology and to the science of human movement is more often than not completely overlooked.

On the score of human movement this is not entirely surprising. First, the laws of mechanics are more easily applied to rigid than nonrigid bodies such as the human body. Secondly, a large amount of the work carried out in this field has been written up in journals either not known or not easily accessible to staff and students in schools.

This small book has been written to help overcome this situation and if by reading it, sixth-formers and other students studying advanced physics courses find the topic of mechanics a little more interesting and relevant, the book will have succeeded in its aim.

A study of the mechanics of human movement can be divided first into the analysis of such movement and secondly into the measurement of the mechanical parameters involved in these analyses. In this book both aspects will be considered. A basic knowledge of mechanics has been assumed so that the derivation of basic equations has been omitted unless they have particular relevance to the topic being discussed. A list of references is included which indicates the sources of information that should be consulted if the reader wants to take matters further. Finally, some suggestions for projects and investigations are included in the hope that the reader will be able to follow up this introduction to the physics of human movement by some practical work of his own.

1 | Velocity analysis of human movement

Meaning of velocity

When an athlete is running in a race, the rate at which he is covering his ground gives a measure of his velocity. If for example he covers 12 metres in 1·5 seconds, we say that he has an average velocity (or coverage rate) of 8 metres per second. Thus his average velocity is defined by the equation

$$\text{average velocity} = \frac{\text{distance covered}}{\text{time taken}} = \frac{s}{t} \text{ m s}^{-1}$$

where s is the distance covered and t is the time taken. There is, however, another type of velocity, called the instantaneous velocity. This is the velocity that, say, a runner has at a particular time. Suppose a runner has an instantaneous velocity of 8 m s^{-1} at 2 seconds from the start of a race. This means that if his instantaneous velocity does not change, he will cover 8 metres in the next second. However, suppose we find at 3 seconds it has changed and he then has an instantaneous velocity of 10 m s^{-1} and has covered 9 metres during the previous second. His average velocity over this period is 9 m s^{-1} and this is not equal to either the instantaneous velocity at 2 seconds or 3 seconds. So instantaneous velocities are not necessarily equal to average velocities.

Constant instantaneous velocity

In fig 1.1(a) the instantaneous velocity at any time is the same, and must necessarily be equal to the average velocity.

$$\text{Average velocity of walker} = \tfrac{10}{5}$$
$$= 2 \text{ m s}^{-1}$$

Instantaneous velocity at 2 seconds, say
$$= 2 \text{ m s}^{-1}$$

Note also that the area under the graph is equal to the distance travelled, a fact true of all velocity-time graphs.

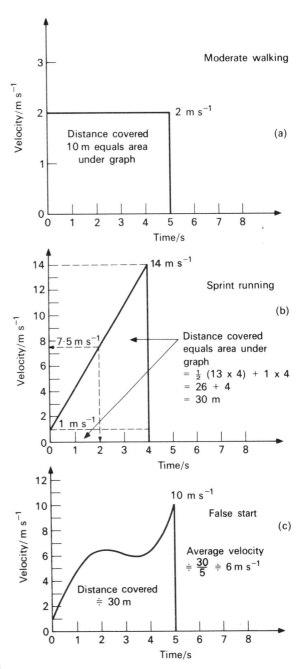

FIG. 1.1 *Velocity-time graphs.*

Uniformly increasing instantaneous velocity

In fig. 1.1(b) each second the instantaneous velocity increases by the same amount. When this happens the average velocity is equal to the average of the initial instantaneous velocity, (u) and the final instantaneous velocity (v). This can be shown as follows.

Distance travelled by runner = area under graph
$$= 30 \text{ m}$$

Average velocity of runner = $\frac{30}{4} = 7.5 \text{ m s}^{-1}$

Initial instantaneous velocity of runner = 1 m s^{-1}
Final instantaneous velocity of runner = 14 m s^{-1} } average

$$= \frac{14 + 1}{2}$$
$$= 7.5 \text{ m s}^{-1}$$

Thus the average velocity = $\frac{u + v}{2} \text{ m s}^{-1}$

It is also equal to the instantaneous velocity at half-time as the reader can confirm: that is at 2 seconds.

Fluctuating instantaneous velocity

In fig. 1.1(c) the instantaneous velocity does not increase regularly and as a consequence there is no significant relationship between the average velocity of the movement and the initial and final values of the instantaneous velocity.

Average velocity of runner = $\frac{30}{5} = 6 \text{ m s}^{-1}$

Initial instantaneous velocity of runner = 1 m s^{-1}

Final instantaneous velocity of runner = 10 m s^{-1}

In future for ease, instantaneous velocity will be referred to simply as velocity while average velocity will be given its full title.

Acceleration

When a runner is increasing or decreasing his velocity, he is said to be accelerating or decelerating. The rate at which this occurs gives a value for his acceleration or deceleration. It follows for the walker of fig. 1.1(a), that the acceleration is zero while for the runner of fig. 1.1(b), the acceleration is constant.

The instantaneous acceleration at any particular time under these circumstances is equal to the average acceleration and both are given by the equation:

$$\text{acceleration} = \frac{\text{increase in velocity}}{\text{time taken}}$$

that is

$$a = \frac{v - u}{t} \text{ m s}^{-2}$$

or

$$v = u + at$$

For the movement of fig. 1.1(c) the acceleration is not constant and the above equation is not valid, except over areas where the acceleration does not change very much. Fortunately many movements can be broken down into stages where this is true and we need take this discussion no further.

Equations governing constant acceleration

From previous sections and in particular the one before last: the equations below have been established to describe movement that possesses constant acceleration.

$$\text{average velocity} = \frac{s}{t}$$

$$\text{average velocity} = \frac{u + v}{2}$$

$$v = u + at$$

By a combination of these equations we get

$$s = \frac{(u+v)t}{2}$$

and

$$s = vt - \tfrac{1}{2}at^2$$
$$s = ut + \tfrac{1}{2}at^2$$

and

$$2as = v^2 - u^2$$

These equations cover the movement of fig. 1.1(b). They also approximately cover vertical free fall such as that experienced in diving, jumping and trampolining. This is because the acceleration due to gravity "g" is constant and equal to 9.8 m sec^{-2}.

Consider the superimposed dive taken from a cine film shown in fig. 1.2. As the initial vertical velocity of the downward part of the dive y is zero and $a = 9.8 \text{ m sec}^{-2}$,

$$s = \tfrac{1}{2}at^2$$
$$= 4.9t^2$$

and thus for equally increasing time intervals from the start of the fall, the distance fallen, s, should increase in the ratio 1:4:9:16. Readers can check this from the figure for themselves. It should also be noted that $v = u + at$ becomes $v = 9.8t$ and thus as t is approximately one second, the entry velocity of the diver is about 10 m s^{-1}. This is just below a sprinter's maximum velocity as a top class sprinter can reach 13 m s^{-1} in about 2–3 seconds during a 100 m sprint.

Moving and turning

So far translational movement, that is movement from one point to another, has been discussed. Most movement, however, consists of a combination of translational movement and rotational movement. Rotational velocity and acceleration can be treated in the same way as we have treated translational velocity and acceleration, except that distance is replaced by revolutions or the angle turned through in radians or degrees. In one revolution there are 6·3 radians or 360 degrees.

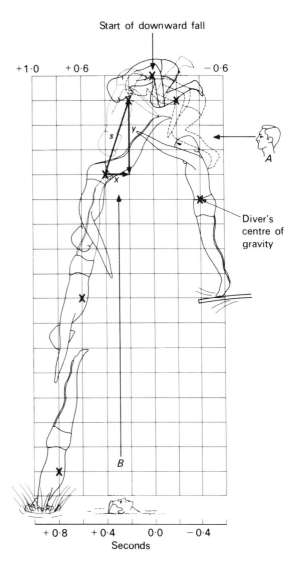

FIG. 1.2 *A diagram of a stroboscopic photograph of a diver taken at 0·2 s intervals.*

Movement in a plane

Not all translational movement takes place in a straight line: some such as a diver or a gymnast trampolining or a jumper high jumping takes place in a plane. It is sometimes convenient to discuss separately the vertical and horizontal components of velocity. Suppose the diver shown in fig. 1.2 moves a distance s metres in reality, during a small time interval of t seconds. An observer at A will see him move y metres while an observer floating on his back at B will see him move x metres. The horizontal component of velocity v_x m s^{-1} will be given by x/t, the vertical component by y/t m s^{-1}, and the actual velocity v by s/t m s^{-1}

But from geometrical considerations

$$s^2 = x^2 + y^2$$

or

$$\left(\frac{s}{t}\right)^2 = \left(\frac{x}{t}\right)^2 + \left(\frac{y}{t}\right)^2$$

so that

$$v^2 = v_x^2 + v_y^2$$

or

$$v = \sqrt{v_x^2 + v_y^2}$$

It can also be shown that as $x = s\cos\theta$ and $y = s\sin\theta$ that

$$v_x = v\cos\theta$$
$$v_y = v\sin\theta$$

so that the direction of the actual velocity to the horizontal component which is defined by the angle θ is given by:

$$\tan\theta = \frac{v_y}{v_x}$$

Methods of measuring human velocities

There are several different ways in which velocity-time graphs can be obtained and each one has its own particular advantages (and disadvantages). The four methods described briefly in this chapter are:

(1) The ticker-tape method,
(2) The cine film method,
(3) The stroboscopic method,
(4) The photo-electric cell method.

The ticker-tape method

The reader is undoubtedly familar with the a.c. tape vibrator, which, when actuated by a 6 or 12 volt a.c. supply, marks a tape fed through the vibrator with a dot every 1/50th of a second. A suitable period to measure distance travelled, is either 1/10th or 1/5th of a second, i.e. over 5 or 10 consecutive spaces. Let us consider 5 spaces:

Suppose the distance over 5 spaces is "s" cm.

$$\text{Time taken} = 5 \times \tfrac{1}{50} = \tfrac{1}{10}\,\text{s}$$

$$\text{Average velocity} = \text{Distance covered} \div \text{Time taken}$$

$$= 10\,s\,\text{cm s}^{-1}$$

This method of velocity analysis is limited in its scope, but can be employed for such things as the sprint start, the long jump run-up, weight lifting and swimming, provided the tape is waterproofed in the latter instance. Fig. 1.3 illustrates the apparatus being used in a spring start. The reasons for its limitations are that the apparatus can only measure linear velocities and is only accurate if the payout of tape is no more than about 30 metres as it then begins to sag appreciably and break. However, within its limitations it is cheap to operate and is reasonably accurate and the results are easy to analyse. The rest of this section will be concerned with an investigation into the sprint start that used this method of velocity analysis.

The ticker tape apparatus was placed on a platform about 1 metre off the ground, so that when the tape was attached to the back of the sprinter's waist it was pulled into a horizontal position

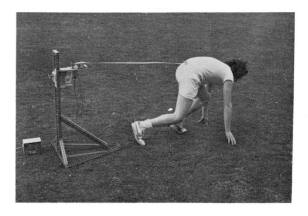

FIG. 1.3 *Ticker tape analysis of a sprint start.*

when the sprinter was fully out of the blocks, i.e. his centre of gravity had reached full height. The vertical movement of the tape from the start is not very large compared with the horizontal distance moved to when the sprinter reaches maximum speed, so that the tape can be considered as measuring the horizontal velocity of the sprinter. Only 25 metres of tape was used as it had been found that maximum speed was usually reached in this distance. However, to simulate competition conditions, the sprinter ran the full 100 metres, was paced by one or two other sprinters and was started by a gun. It is best for the tape to be wound on to a spool with an arm brake which controls the length of tape being paid out.

The tapes obtained show two places where several dots were superimposed. The second was taken as the start as the first represented the "set" position and the distance between the two sets of dots, the distance the sprinter's back moved from the "set" to "go" positions. The blocks were placed about 1 metre in front of the ticker tape apparatus. Fig. 1.4 represents one set of results gained during the investigation using a 40 cm block spacing for an 18 year old student. A velocity-time graph has been plotted from data obtained from the tape. Note that the velocities calculated are average velocities for consecutive 0.2 second intervals. This is shown by the step graph. The velocity curve drawn as a heavy black line is gained by joining the mid-point of each step together. That is, the step velocity between 0.4 and 0.6 seconds is the average velocity over this period and, assuming reasonably constant acceleration over the period, it is also the instantaneous velocity at 0.5 seconds.

A lot of useful information can be gained from such a graph.

(a) *Maximum Velocity*

This is marked on the graph as V_m and corresponds to a velocity of 9.15 m s^{-1}. The time to reach this velocity can be read off the time axis and the graph in question gives a value of 2.8 seconds. It might be argued that as the tape has almost run out at this point there may be a second increase in velocity later on. A cine film analysis indicates that there is sometimes a burst of speed near the end of the 100 m, but this is offset frequently by the runner losing slightly on the maximum velocity to which he builds up after 2 to 3 seconds. In the case in question the distance covered in reaching maximum velocity is 18.84 m. This leaves 81.16 m still to be covered.

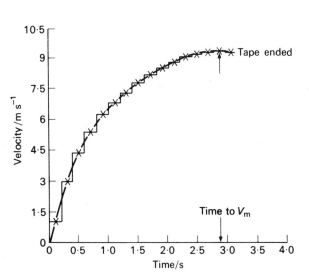

FIG. 1.4 *Velocity-time graph of a sprint start.*

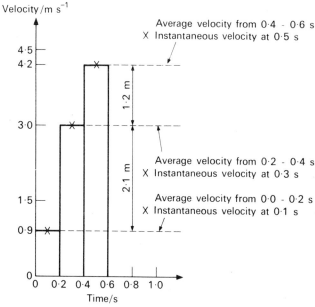

FIG. 1.5 *Using a velocity-time graph to calculate accelerations.*

Assuming the runner maintains his maximum velocity at 9.15 m s^{-1}, this would take 8·9 s, giving him a total of 11·7 s. The stop-watch used as a check in the project clocked 12·1 s which would indicate some slowing down.

(b) *Acceleration-Time Graph*

This involves finding the slope of or differentiating the velocity-time graph. Probably the easiest way of doing this is as follows. Suppose during the first 1/5th second the average velocity is v_1 and during the next 1/5th of a second it is v_2. The increase in velocity is $(v_2 - v_1)$ and the time taken for this to happen is 1/5th of a second. Thus, the acceleration is given by:

$$\text{Acceleration} = (v_2 - v_1) \div \tfrac{1}{5}$$
$$= 5(v_2 - v_1) \text{ m s}^{-2}$$

Consider the example represented in fig. 1.5.

Average velocity for the first
fifth of a second = 0.9 m s^{-1}

Average velocity for the second
fifth of a second = 3.0 m s^{-1}

Acceleration from first to second
fifth of a second = $5(3.0 - 0.9)$
$= 10.5 \text{ m s}^{-1}$
= instantaneous acceleration at 0·2 secs

This acceleration represents the instantaneous acceleration at 0·2 seconds which in this instance is greater than the acceleration due to gravity, g!

Thus, an acceleration-time graph can be drawn up from the velocity-time graph and Table 1 has been drawn up in this way from the smoothed graph of fig. 1.4.

(c) *Average Horizontal Force*

By Newton's second law the average horizontal force producing acceleration will be given by:

Average Horizontal Force = Sprinter's Mass
× Acceleration

Table 1

Time in seconds	0·2	0·8	1·4	2·0	2·6	3·0
Acceleration in m s^{-2}	10·5	3·8	2·1	1·2	0·6	0·0
Force as fraction of body weight	1·07	0·39	0·21	0·12	0·06	0·0

Table 1 shows the results gained for the sprinter we have been considering whose mass was 77·3 kg, and weight 757 N.

It may raise some comment that a runner is apparently exerting no net horizontal force at maximum velocity. In practice the force being determined above is the resultant of horizontal force exerted by runner minus the air resistance experienced by him, so that when maximum velocity is reached the runner still has to exert an average horizontal force to overcome air resistance. Also, the style of many runners has a small decelerating phase and accelerating phase which combined together give a net acceleration of zero.

(d) *Comparison of Performance*

The best way to compare performances of different types of start, and between different sprinters, is to compare distance-time graphs produced with a ticker tape.

This analysis technique has been used by Dr Hoffman[1] at research level in investigations on the long jump and triple jump. The apparatus was more sophisticated and nylon thread was used instead of tape. The marks were placed on the thread by an ink marker. A similar piece of equipment is also in use at the Biomechanics Laboratory of San Francisco Veterans Hospital. It has also been used by Brodie and Woolnough at Abingdon School to compare the performance of weight lifters.[2]

Cine film methods

This method often goes under the heading of kinegraphic analysis and before any projects are described illustrating this technique some comments about the technique and the apparatus involved must be made.

(a) Cine Cameras

Electrically-driven cameras in general tend to have a more constant frame speed than clockwork-driven cameras. Also, the more frames per second that are taken, the better the definition. This becomes very important when fairly high velocities are involved, such as those that are achieved in the tennis serve, the golf drive and the javelin throw. The best results are, therefore, achieved with at least 64 frames a second, although results can be gained from 16 frames per second cameras for slow activities.

(b) Analysis and Display

If a descriptive analysis is being carried out, superimposed outlines of selected frames on a white sheet of paper are usually sufficient. To distinguish the order of the positions it is often helpful to use a colour code. For example: position one—red, position two—blue, position three—green, and so on. If a quantitative analysis is required, it is sometimes more useful to make the outlines on an acetate sheet marked with a grid especially if a centre of gravity analysis is required. This often is the case as the point of interest is the movement of the centre of gravity of the performer.

(c) Errors

(i) PERSPECTIVE ERRORS Unless special account is being taken of these errors, the angle of view should be restricted to 25°, otherwise the scale factor will change from the extremes to the centre of the movement by more than a negligible amount.

(ii) BUCKLING Check the vertical lines of any apparatus or building for this error which only occurs with very cheap cameras.

(iii) FRAME SPEED This will always tend to vary and therefore in accurate work the frame speed marked on the camera is not a reliable indication of time intervals. Placing a large continuous movement clock in the field of view is the best answer to this problem.

Cine film analysis can be separated into two categories—analysis when there is large scale linear movement and analysis when there is little linear movement but a large amount of rotational movement.

(d) Cine Analysis with Large and Small Linear Movement

To reduce perspective errors where large linear movement is involved the camera ought to be placed on a perpendicular line from the mid-point

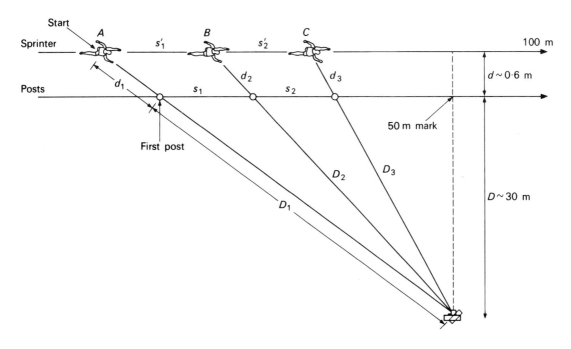

FIG. 1.6 *Filming a 100 m sprint.*

Table 2

Distance covered by the runner m	1·55	3·11	6·22	9·33	12·44	15·55	18·66	21·77	24·88	27·99	31·10	40·43	49·76	59·09	68·42	77·75	87·08	96·41	100·00
Number of frames	27	17·5	28	25	23·5	22	22	21·5	22	22	21·5	66	67	67	67	68	66	63	24·5
Average velocity m s^{-1}	3·7	5·7	7·1	8·0	8·5	9·0	9·0	9·3	9·0	9·0	9·3	9·0	8·9	8·9	8·9	8·8	9·0	9·5	9·4

Camera speed = 64 frames/second. Correction factor = $(1 + d/D)$ = 1·02.
Total distance covered by runner = 100 m. Total of 716 frames giving a total time of $716 \times \frac{1}{64}$ = 11·2 seconds.

of the movement, at a distance of at least one-third the linear distance being filmed. Take for example the 100 m sprint shown diagrammatically in fig. 1.6. The distance "D" should be at least 100 metres. Distance markers are necessary as the camera cannot be kept stationary, and these can be posts distinguished by a white triangle of cardboard attached to the top of each one of them. The camera should always be pointed at the centre of the runner and to help in this direction it is best to fit crosswires over the view finder and follow the runner with these. The runner is then filmed along a path represented in the diagrammatic figure (fig. 1.6) by s'_1, s'_2, etc. The camera needs to be set on a firmly-based tripod to do this successfully.

When the film is analysed either by editor or by projection on to a screen, whenever the runner is intersected by a post, he must be at one of the points marked on the diagram by A, B, C, and so on. It follows that:

$$s'_1 = s_1 \left(1 + \frac{d}{D}\right)$$

$$s'_2 = s_2 \left(1 + \frac{d}{D}\right)$$

where s_1, s_2 etc. are the distances between the posts, D is the distance the camera is set back from the post at the mid-point of the run, d is the distance the runner runs behind the posts, and s'_1 and s'_2 etc. are the distances covered by the runner. Thus the bracketed term is a standard correction and the average velocity is given by:

$$\text{Average velocity} = \frac{\text{Distance between posts} \times \text{standard correction} \times \text{frame speed}}{\text{Number of frames counted between posts}}$$

Table 2 gives the results gained for a 20 year old College of Education student with the arrangement as in fig. 1.6. The stop-watch gave 11·1 seconds. One pacer was used wearing a dark vest to prevent confusion on analysis. The film was taken in black and white. From these results it can be seen that maximum velocity was gained about 20–23 m and 95–100 m.

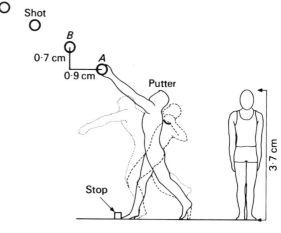

FIG. 1.7 *Analysing a shot-put.*

Not all translational motion is linear of course. Take for example the shot put. Here the motion is two dimensional. It is still possible to calculate the vertical and horizontal components of velocity by the technique just outlined. Fig. 1.7 shows a superimposed trace of an 8 mm regular film which indicates the position of the shot at 1/16th second intervals. Before the put was attempted the student was filmed standing still to enable the scale to be calculated for subsequent calculations. His full height was 1.8 metres and on the trace 37 mm. Thus 1 cm on the trace represents 0·486 metre in reality. The shot left the hand at A.

Distance travelled from A to B vertically,
"y" say $= 0.486 \times 0.7 = 0.34$ m.

Distance travelled from A to B horizontally,
"x" say $= 0.486 \times 0.9 = 0.44$ m.

Assuming wind resistance is negligible at this early stage of the motion, the horizontal velocity v_x will be constant. Therefore:

$$v_x = \frac{x}{t}$$

The vertical motion will be affected by the pull of gravity. Thus the vertical velocity at projection will be gained by substitution in $s = ut + \frac{1}{2}at^2$ remembering $a = -9.8$ m s^{-2}. Thus

$$y = v_y t - 4.9 t^2$$

or

$$v_y = \frac{y}{t} - 4.9 t^2$$

Therefore with $t = 1/16$th second

$$v_x = 7.0 \text{ m s}^{-1}$$
$$v_y = 5.5 \text{ m s}^{-1}$$

The angle of projection $= \tan^{-1} \dfrac{v_y}{v_x}$

$$= 39° \ 24'$$

and the velocity of projection $= \sqrt{v_x^2 + v_y^2}$

$$= 8.9 \text{ m s}^{-1}.$$

For maximum range this angle is probably a little high. This may seem a little confusing, but the angle of 45° frequently quoted only applies to ground to ground projections where the velocity of projection is independent of the angle of projection. The shot putter satisfies neither of these conditions and his optimum angle lies between 38° to 40°. If the camera is set far enough back to take the whole flight, it is possible to compute the landing velocity and using these two velocities calculate the range of the shot. This can then be checked against the measured range as a guide to accuracy. On the above trace the full shot put range was not recorded but when checks have been made on other films, the accuracy has been within 2–3%.

(e) *Cine Analysis Where Motion Is Mainly Rotational*

These two sub-divisions are very loosely drawn as the technique for the two cases to be discussed under this heading is exactly the same as that used for the last two cases of the previous heading. The only distinction between them is that in the two cases to be discussed the angular rather than linear velocities are the main consideration. The first type of angular motion is illustrated by the Grand Circle in gymnastics where the whole motion takes place about a fixed axis. Examples of this type of motion are few and are confined to gymnastic activities. The camera must be securely fixed and its line of view at right angles to the movement and far enough away to get the full range of movement in its field. Superimposed traces are then made and the scale ratio calculated from some known length such as the performer hanging full length from the bar. As the force developed by the arms can be calculated from this velocity analysis it will be more fully discussed later.

The second case to be discussed under this heading involves both linear and rotational movement and the latter has to be separated from the former. There are several instances of this type of motion but we shall use the back somersault as our example. Fig. 1.8(a) shows a student performing this movement at $\frac{1}{4}$ second time intervals.

The translational velocity at any point can be calculated in the same way that it was calculated for the shot-put using the distances between the position marking on the centre of gravity and the vertical or horizontal so that these slides can be superimposed to produce an outline like the one shown in fig. 1.8(b).

Using the line joining the centre of gravity to the shoulder as the "rotating" line, the angular velocity of the shoulder about the centre of gravity is given by:

$$\text{angular velocity} = \frac{\text{angle moved through by rotating line}}{\text{time taken}}$$

Stroboscopic method

This method is really an extension of the previous one but it does allow a more detailed analysis to be carried out because smaller time intervals can be used. It also superimposes images directly which is an advantage, but it cannot be used where there is large translational movement as the camera must remain fixed.

(a) *Xenon Lamp Method*

A xenon stroboscopic lamp is set up to illuminate the subject and is set at a suitable flashing time, such as 1800 flashes a minute. Each flash lasts 10 μs and gives minimal illumination so that a high speed film is necessary. A Land polaroid camera, Type 180, which uses a 3000 speed film is a suitable camera. It also has the added advantage of producing a fully developed film in 15 to 20 seconds. The shutter speed has to be set to manual operation (the B setting) and is operated to cover the subject's movement. The stop number usually needs to be set to 4·5. As before, the camera must be at right angles to the plane of movement and must be rigidly fixed. This means using a very good tripod. To get the scale multiplication factor a feature in the plane of movement must be filmed and measured in reality and on the film.

This method has the great disadvantage of low intensity and pulsating illumination which distracts the subject and prevents such actions as the golf drive from being filmed unless a bank of stroboscopic lamps is used. Fig. 1.9 shows a golf stroke photographed by this method using a bank of strobes and the reader is left to analyse this picture for himself given that the golf club

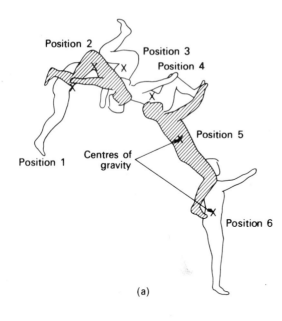
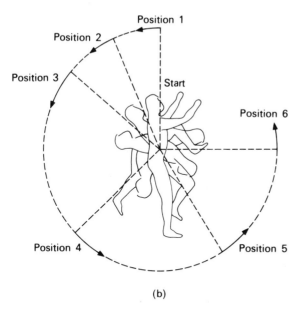

FIG. 1.8 *Estimating the rotational velocity of a back somersault.*

centre of gravity positions to determine the velocities in the vertical and horizontal directions. To find the rotational velocity the centres of gravity must be superimposed. To do this properly it is best to make individual acetate slides of each

VELOCITY ANALYSIS OF HUMAN MOVEMENT

FIG. 1.9 *A stroboscopic golf drive.*

FIG. 1.10 *A stroboscopic tennis serve.*

shaft in the bottom six positions measures 1 metre in reality and the strobes were flashing 6000 flashes per minute. This method, however, does create little blur even for high speeds because of the short time interval of the flash, as blur depends on the movement, during an exposure, of the subject being photographed, produced on the film itself. High speed films are more tolerant on this score than low speed. Thus less blur is obtained from Polaroid work than cine-film work. Blur can be reduced by either shortening the exposure time and/or moving the subject further from the camera so the scaling factor is increased. With cheaper cine cameras and still cameras only the latter can be altered, and as these cameras are often used with low speed film a certain amount of blur has to be tolerated.

(b) *Stroboscopic Disc Method*

The disadvantages of the above method are overcome with a stroboscopic disc set up in front of the camera. The subject can be constantly illuminated with floodlights so that any movement can be filmed.

The slits on the disc rotating in front of the camera allow a flash of light to enter the camera of longer duration and higher intensity than the xenon lamp. It should be remembered that small slits require higher illumination but lessen the blur. Depending on the number of floods used and the slit width, the stop number can vary from 4·5 to 16.

Fig. 1.10 was taken by the BBC using this technique.

Photo-electric cell method

This method allows very high speeds such as those gained by a golf club to be measured in the school context. Two photo-electric cells are used to stop and start a millisecond timer. A suitable arrangement is shown in fig. 1.11.

The timer, gating unit and photo-electric cells are made by Panax Ltd. When the beam is cut between the first light bulb and cell, the timer starts and when the beam between the second

FIG. 1.11 *Photo-electric cells being used to measure the velocity of a golf drive.*

light bulb and cell is cut, the timer stops. With such an arrangement one of the author's past students has investigated the influence of the club head mass on the drive given to a golf ball. This study was based on work originally published by Daish.[3] The golf club used is shown in fig. 1.11; discs were screwed to the end of the shaft to gain the different club head masses and the experiment was conducted indoors with a ping pong ball taking the place of a golf ball. The results confirmed Daish's work but with a slightly wider range of power coefficient. If a gating unit and cells are not available, the arrangement of fig. 1.12 can be used. When the first wire is broken the timer starts, and stops when the second wire is broken.

FIG. 1.13 *One of the first pictures taken of human and animal movement.*

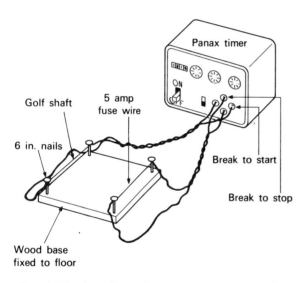

FIG. 1.12 *An alternative arrangement to photo cells.*

FIG. 1.14 *The arrangement used by Muybridge to take the first pictures of human and animal movement.*

Because of the inconsistent stretching of the wire before breaking, results using this set-up are not as reliable or consistent. However, Muybridge took the first photographs of animal and human movement by using a technique similar to this except that as his wires were broken they fired cameras as fig. 1.14 shows.

With more expensive photo-electric cells a larger separation can be achieved and then this device can be used to measure the speed of an athlete at a given distance and so on. With a bank of cells it is possible to obtain velocity-time graphs, but this is an expensive procedure. An alternative, on which the author is working, is a bank of closed copper wire loops relayed back to an ultra-violet recorder. A magnet is strapped to the back of the sprinter and every time he runs through one of the loops, an induced e.m.f. is produced in the loop, which the ultra-violet recorder picks up and records against time. Average velocities can be obtained by knowing the separation of the loops and measuring the time intervals between the peaks on the ultra-violet recorder. It may be possible to measure instantaneous velocities as the height of the peaks is a function of the speed the sprinter and magnet passes through the loop. As yet this project is only in its initial stages.

2 | Human forces

Introduction

In the last chapter we were concerned with describing movement in terms of velocity and acceleration. In this chapter we shall be more concerned with the forces that cause movement. Basically forces can produce three types of movement:

(1) deformative movement—where the force changes the shape of the body

(2) translational movement—where the force moves the body from one place to another

(3) rotational movement—where the force rotates the body.

Very often, more than one type of movement results from the application of a force. One example that usefully illustrates this point is the corner kick illustrated in fig. 2.1. The force the ball receives from the player as well as giving it a spin and sending it forwards, also deforms it.

There are several types of force. All of the principle forces that are experienced in human movement result from contact either with the ground, or the water, or the air, with the exception of the force of gravity. When a high jumper leaves the ground he is immediately pulled back towards the ground so that his upward movement is decelerated till it is reversed to an accelerated downward movement. The force that causes this is known as the weight of jumper, and it is the force due to gravity. The point at which this force appears to act is called the centre of gravity, and we shall be discussing this parameter more fully in the next chapter.

When he reaches the ground the jumper's fall is halted. This does not mean that the force of gravity has ceased and that the jumper has no weight. Rather, another type of force has come into play due to the jumper's contact with the ground. This type of contact force is called a normal contact force, and this force is experienced whenever a body is in contact with the ground. Since there is no vertical movement the weight (W) of the jumper must be balanced by the normal contact force (N). This is illustrated by using arrows of equal length. The normal contact force always acts at right angles to the surface in contact as shown in fig. 2.2(c) and (d).

(a) Force of gravity decelerating jumper

(b) Normal contact force opposing jumper's weight as he lands

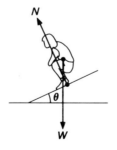

(c) Normal contact force

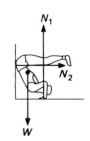

(d) Two normal contact forces

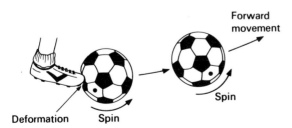

FIG. 2.1 *The action of a footballer's kick on a football; deformation, spin and translation.*

FIG. 2.2 *Types of force acting on body in human movement.*

Another important force that we need to consider is the force of friction. When we rub our hands along a bench, we can feel the movement being resisted by this particular force. As frictional forces play a very important part in human movement we shall discuss them a little more fully in the next section.

Frictional forces

As a frictional force always opposes the relative motion of two surfaces, it can sometimes hinder movement and be an unwanted effect. For example as fluid friction, in the form of drag resistance, it hinders the movement of a swimmer through the water and to a lesser degree a diver or high-jumper through the air.

At other times, however, it helps movement and becomes a very vital force. For example, consider the foot action of someone walking along a path, which is an example of contact friction. The muscles of the leg push back against the path and this is matched by the force of friction set up between the foot and path. As a result it is not the foot that moves but the walker's body instead. If the force of friction is reduced, as it is if the path is icy, the foot may not be held firm and slips back. Then locomotion becomes very difficult indeed.

Let us examine two forms of frictional force a little further.

Contact friction

This type of friction can be more fully investigated with the apparatus shown in fig. 2.3. The load produced by the slotted discs (P) will be opposed by the force of friction (F) until the block starts to move, when F will have reached its maximum value (F_m). Therefore at the point of slipping

$$P = F_m$$

and investigation shows that when this occurs, irrespective of the size of the block,

$$F_m = \mu_s N$$
$$F_m = \mu_s W$$

as N must be equal to W. The coefficient μ_s is called the coefficient of "static" friction and is a

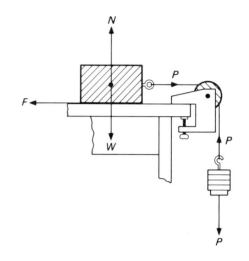

FIG. 2.3 *Measurement of coefficient of slipping friction.*

constant that is determined by the two surfaces in contact. When the block is actually moving, the force of friction drops slightly, but it is still governed by a similar relationship, namely:

$$F = \mu_d N$$

where μ_d is called the coefficient of "dynamic" friction. "μ_d" is, of course, less than "μ_s". Table 3 gives the values of μ_s for some surfaces which come into contact in human movement studies.

Table 3

Surface in Contact	μ_s
Nylon on nylon	0·15–0·25
Rubber on solids	0·75–4
Wood on wood	0·25–0·5
Leather on wood	0·6
Leather on cork	0·5

It follows that the larger the value μ_s has for two surfaces, the better the grip. This is very important in the manufacture of such things as basketball shoes, where the better the grip the greater the lean a player can have before he slips over.

Consider the player of fig. 2.4. As he pushes into the ground with a thrust (T) it will lie along his body line. The contact force (N) and frictional

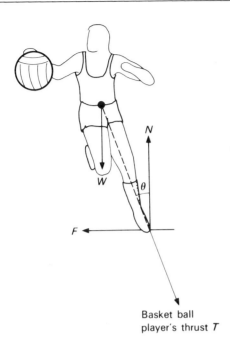

FIG. 2.4 *Angle of lean.*

force (F) will oppose this thrust and therefore be equal and opposite to it. For this condition to be satisfied: resolving horizontally gives us:

$$F = T \sin \theta$$

and vertically:

$$N = T \cos \theta$$

Thus

$$\tan \theta = \frac{F}{N}$$

If the player is about to slip, θ will equal θ_m the maximum angle of lean, and F will equal F_m which in turn will equal $\mu_s N$, so that the equation becomes:

$$\tan \theta_m = \mu_s$$

Thus, if the coefficient of friction μ_s for rubber shoes on a maple floor is 0.75, the angle of lean for a player in this case would be given by:

$$\tan \theta_m = 0.75$$

or

$$\theta_m = 37°$$

Drag friction

Whenever a swimmer moves through the water or an athlete jumps through the air, they experience a resistive force called a drag force. The nature of this force is not easy to explain but it does depend on the effective cross-sectional area that is presented by the swimmer or jumper at right-angles to the direction of their movement, their speed, the "streamline" of their body configuration, and the density of the medium through which they are travelling. Dimensionally it can be shown that the dependence of the drag force on these factors must have the following form:

$$F_D = K \rho A v^2$$

where ρ is the density of the air or water, A is the effective cross-sectional area presented by the jumper or swimmer and v the velocity of either of them. K is a constant dependent on the "streamline" of their body posture and obviously as neither K, A nor v remain constant for either the swimmer or the jumper, the drag force they experience is a continually changing quantity.

Force and linear acceleration

Consider a long board laid on a set of rollers. If it receives a force to the right it will move to the right, whereas if it receives one to the left it will move to the left. If it receives no net force at all it will remain stationary. When a runner crosses the board at a steady velocity we find no net movement of the board. It only moves when he accelerates or decelerates on the board. Thus forces such as those the runner gains from the ground are used to produce either acceleration or deceleration. At steady velocity very little force is required and that is only needed to overcome frictional effects. The magnitude of the acceleration caused by a force will of course depend on the mass of the body being moved. For a given force, the larger the mass the smaller the acceleration. This is why sprinters are lightweights.

Impulse and momentum

A force rarely acts instantaneously and we therefore find that for the football kicked in fig.

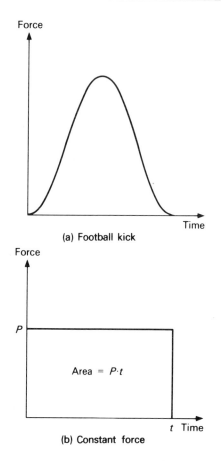

FIG. 2.5 *Force-time graphs.*

2.1 the force experienced by the ball changes with time as the force-time graph of fig. 2.5(a) shows.

Suppose for convenience a force P acts for a time t on a football of mass m, producing an acceleration "a" which changes its velocity from u to v (see fig. 2.5(b)). We define the product of a force and the time it acts, as the impulse of a force.

$$\text{Impulse} = Pt$$

You will remember that we can show that the impulse is equal to the change of momentum that it causes and

$$\text{Impulse} = \text{Change in momentum.}$$

This is a very important equation and true for varying forces as well as fixed ones. For example the force platforms to be discussed later measure impulse rather than force, therefore this equation is vital in relating the impulse developed by a performer to the changes in his momentum.

For the case just considered note that the area under the graph is equal to Pt and thus is equal to the impulse and the change in momentum this impulse causes. This is also true for any force-time curve, so it would be true of fig. 2.5(a) as well.

Another important aspect of the equation comes out when we consider two bodies colliding. When two balls collide the impulse received by ball A from ball B (I_2) is equal and opposite to that received by ball B from ball A (I_1) so that

$$\frac{\text{Change in momentum}}{\text{of ball } A} = \frac{\text{Change in momentum}}{\text{of ball } B}$$

which means that the linear momentum of the system is conserved. This is an important principle which explains many phenomena, and has many applications. It is interesting to consider the implications of this principle of conservation of linear momentum when a bat or club hits a ball, and also when a ball is thrown or caught.

Rotational effect of forces

Anyone who has sat on a see-saw has experienced the turning effect of a force, and the lighter their partner has been the further away from the centre he has had to sit to balance them. This leads us to define the turning effect of a force in terms of its moment which is the product of the force and the perpendicular distance from its line of action to the pivot.

If there is no fixed pivot for a body its centre of gravity acts as the pivot or axis. Thus if a football is given a side blow F_1 by introducing two imaginary forces F_2 and F_3 equal in magnitude to F_1 at the centre of gravity of the football we can see that F_1 and F_2 produce a couple of magnitude:

$$G_1 = F_1 x_1$$

which rotates the ball, while F_3 moves the ball translationally. The introduction of these two imaginary forces does not change the situation, as the resultant of them is zero.

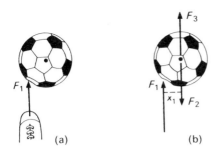

FIG. 2.6 *A football kick.*

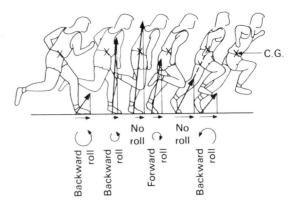

FIG. 2.8 *The "roll" of a runner during one landing and one take-off.*

These ideas can probably be better understood by considering their application to a specific human movement situation such as the sprinter or the diver.

Fig. 2.7 shows the two forces acting on a sprinter, namely his weight W and the thrust he evokes from the ground T_1. By introducing two opposite forces T_2 and T_3 at his centre of gravity equal in magnitude to T_1 we can see that the couple G_1 (of magnitude $T_1 x$) rotates his body backwards, while the force R (the resultant of W and T_3) moves his body forwards and upwards. Only if T_1 passes through the sprinter's centre of gravity will his body not be rotated, for x_1 will then be zero. Fig. 2.8 shows us that this only happens twice during the sprinter's stride. On landing the sprinter's thrust moves from in front of his centre of gravity to behind it, and then moves in front of it again just before he takes off. As a consequence the sprinter's body tends to be rotated backwards, then forwards and then backwards in any one landing and take-off.

In practice the sprinter avoids this "rolling" effect by moving his arms and free leg in such a way as to counterbalance it. How such arm and leg actions do this will be discussed in more detail later. Fig. 2.9 shows the two forces acting

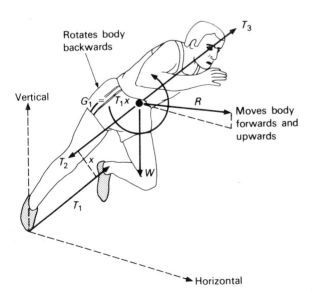

FIG. 2.7 *The sprinter.*

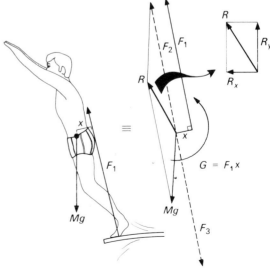

FIG. 2.9 *Diver gaining impulse from the board to translate and rotate himself into his dive.*

on a diver, namely his weight Mg and the thrust he experiences from the board F_1. In the equivalent force system (F_1, F_2, F_3) it becomes clear that F_1 and F_2 link together to produce a couple G of magnitude $F_1 x$ that rotates the diver forwards into his dive. Force F_2 on the other hand combines with his weight Mg to give a resultant translational force R. It is then possible to break R down into a horizontal component R_x which moves the diver horizontally forwards and a vertical component R_y which moves him vertically upwards. R_y is frequently called the normal contact force and R_x is, of course, the frictional force developed between the diver's feet and the diving board.

Circular motion

If a steel ball is released down a runway it will continue in a straight path unless it passes a magnet placed at right angles to its path when it will be deflected on to a circular path. This simple experiment illustrates the first important point about circular motion—to move in a circle an object must be pulled inwards by a force. Such a force is called a *centripetal* force.

Consider the hammer thrower swinging the hammer in fig. 2.10.

To an observer the hammer will appear to be accelerating inwards. Let this acceleration be a and the velocity of the hammer be v, and let us study the hammer's motion from A to B. This inward acceleration is, of course, the direct product of the inward force and the two are related by Newton's law

$$F = m \cdot a$$

As the acceleration towards the centre is equal to v^2/r the inward force involved in circular motion is given by

$$F = \frac{mv^2}{r}$$

where m is the mass of the object, i.e. the mass of the hammer in the example we have just been considering.

Let us apply this equation to a specific hammer throwing example. The length of the hammer thrower's arm is $2/3$ m and the hammer wire 1 m while the mass of the hammer is 7 kg. Suppose the hammer thrower lets the hammer go at $20 \, \text{m s}^{-1}$. Then at release

$$F = \frac{7 \times 20 \times 20}{1 \cdot 67} N$$

$$= 1677 \, N$$

Before leaving hammer throwing to look at some other examples of circular motion, let us consider the forces acting on the hammer thrower. If the hammer must be pulled inwards by a centripetal force, the hammer thrower will

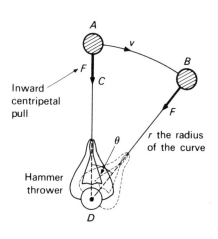

FIG. 2.10 *Circular motion.*

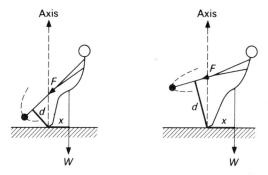

(1) Couple Fd increasing

(2) Lean to produce couple Wx increasing

FIG. 2.11 *The hammer thrower.*

HUMAN FORCES

feel himself being pulled outwards by the so-called *centrifugal* force. Forces always go in pairs like this—one being a reaction to the other. In fact, the phenomenon is frequently referred to as Newton's law of action and reaction. To balance the couple set up by this centrifugal force the hammer thrower has to lean backwards so his weight can provide a counterbalancing couple. As the speed of the hammer increases, so does the centripetal force required by the hammer and the centrifugal force experienced by the thrower. To counterbalance this increasing couple he has to lean further back so that the counterbalancing provided by his weight can increase also.

Human movement contains several examples of circular motion—we shall only consider two of these.

A runner cornering on the flat and on a bank

In order to get the necessary inward force for cornering on the flat a runner has to rely on the force of friction, as shown in fig. 2.12(a). This frictional force arises as he leans over and, in order not to overbalance, the couples set up by the frictional force and the contact force must be equal. Fig. 2.12(b) shows that for this to be the case

$$G_N = G_F$$

so

$$N \cdot a = F \cdot b$$

or

$$\frac{F}{N} = \frac{a}{b}$$

but as

$$\tan \theta = \frac{a}{b}$$

this means that

$$\tan \theta = \frac{F}{N}$$

Fig. 2.12(c) on the other hand shows the translational forces left after the forces setting up the couples G_N and G_F have been considered and removed. From this figure we can see that by resolving vertically

$$N = Mg$$

and horizontally

$$F = \frac{Mv^2}{r}$$

Thus

$$\frac{F}{N} = \frac{v^2}{rg}$$

By putting these equations together, we can eliminate F/N and write:

$$v^2 = rg\tan \theta$$

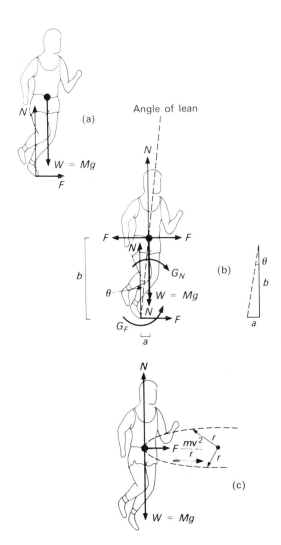

FIG. 2.12 *Cornering on the flat.*

which allows us to predict the lean of a runner for any cornering velocity and cornering radius. If for example velocity is doubled, the radius must be quadrupled for the same angle of lean to apply.

However, there is a limit to θ, for the frictional force cannot be larger than "μN" where μ is the coefficient of friction between the runner's shoes and the track. Thus $\tan \theta$ cannot be larger than μ and so a limit is put on v^2, namely

$$v^2 \leqslant \mu rg$$

or

$$v \leqslant \sqrt{\mu rg}$$

If a runner tries to exceed this velocity for a given cornering radius, he will have to lean over too far to balance and his feet will slip outwards as $G_N > G_F$. On the other hand when cornering on a bank, a runner does not need the force of friction as the horizontal component of the contact force can provide enough inward force by itself as fig. 2.13 shows.

From fig. 2.13 horizontal resolution gives us

$$N \cos(90 - \theta) = \frac{Mv^2}{r}$$

i.e.

$$N \sin \theta = \frac{Mv^2}{r}$$

while vertical resolution gives

$$N \cos \theta = Mg$$

Combining these equations to eliminate N, leaves us with:

$$\tan \theta = \frac{v^2}{rg}$$

or

$$v = \sqrt{rg \tan \theta}$$

where v is the velocity that a runner must have to corner on a bank of radius r and slope θ. If his velocity is larger than this, his feet tend to slip outwards and the resulting frictional force will give the necessary extra inward force needed. If his velocity is less than this the reverse will happen.

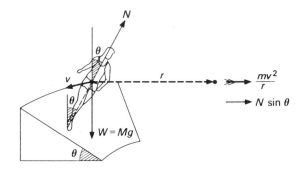

FIG. 2.13 *Runner cornering on a bank.*

Measurement of forces

Having looked at the role forces play in human movement studies, let us have a brief look at the methods available for estimating their size and direction of action.

Indirect measurements

All indirect methods of measuring the force involved in an activity rely on gaining acceleration-time graphs from velocity-time graphs and then using Newton's law to calculate the force, namely

$$F = ma$$

The section in Chapter 1 on the ticker-tape method of gaining velocity-time graphs discussed the basic technique for doing this and actually derived values for the horizontal force for a sprint run (see Table 1).

The limitations of the ticker-tape method as far as sprinting is concerned is the fact that only the *horizontal* component of the total force is measured. Also unless frames of a cine film can be synchronised to specific points on the tape, it is difficult to relate this force to particular body postures.

However, in an activity such as swimming where the net propulsive force is nearly always horizontal and the leg and arm actions are slower and more well defined, these problems are not so dominant. In this case, velocity-time graphs can be produced from water-proofed lengths of tape fed through an a.c. vibrator, one end of which is tied to the swimmer's back. The tape must be kept taut and thus the paying-out spool needs to be effectively braked to prevent the tape sagging. This is even more vital than in the sprint run as the swimmer's motion is more erratic than the sprinter's.

The same indirect analysis from velocity-time graphs can be made from stroboscopic photographs; for example, the stroboscopic photographs of the tennis serve and golf drive in Chapter 1 can be analysed to give a velocity-time graph and then an acceleration-time graph from which the net accelerating force acting on the racket or the golf club can be determined. Cine film analysis can also lead to force-time graphs, but care is needed here for very accurate superimposition must be achieved if the results are to be at all reliable.

In this sort of analysis, the motion of the centre of gravity is of prime interest to the researcher because it is this parameter that acts as the reference point for the translational acceleration and the rotational acceleration that are being produced, as figs 2.7 and 2.9 remind us. Methods of locating the centre of gravity are discussed in the next chapter which is specifically concerned with the centre of gravity of the human body.

One area where the indirect determination of net forces is particularly susceptible to straightforward computation is gymnastic activity carried out on the high bar. Fig. 2.14 shows 8 positions of a gymnast doing a giant swing. His centre of gravity in each position is marked by a dot. By estimating the distance travelled from one position to the next by the gymnast's centre of gravity, it is possible to work out his average velocity between two positions. Then the average inward force T his arms are having to exert can be found by resolving the effect of the forces acting along an inward radius: namely

$$(T - Mg\cos\theta) = \frac{Mv^2}{r}$$

or

$$T = \frac{Mv^2}{r} + Mg\cos\theta$$

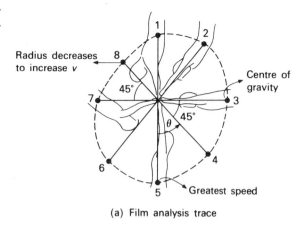

(a) Film analysis trace

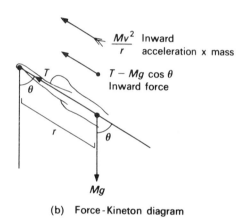

(b) Force-Kineton diagram

FIG. 2.14 *Selected frames of a giant swing together with a force-kineton diagram relating the inward force* $T - Mg\cos\theta$ *to the inward mass × acceleration according to Newton's Second Law* $F = ma$.

where M is the gymnast mass, r the average radius from the bar to his centre of gravity between the two positions, v the average velocity between the two positions, θ the inclination angle to the downward vertical. Table 4 gives the velocities developed by the 73 Kg 1·75 m gymnast and the total arm force needed. Note that the results refer to average rather than specific values of the tension in the arms. The -ve sign indicates that the gymnast has to push himself up rather than pull himself in.

Table 4

Position in swing	Velocity of a giant-swing gymnast					Arm force developed by giant-swing gymnast			
	Time between positions (s)	Distance travelled between positions (cm)	Average radius (cm)	Average velocity (m s^{-1})	Angle of inclination (θ)	Average value of $Mg \cos\theta$ (N)	Average value of Mv^2/r (N) derived from column 4 and 5	Total arm force (N)	As a fraction of body weight (gymnast's weight = 715.4 N)
1					180				
	0.63	176.8	116.4	2.81		−610	494	−116	−0.16
2					135				
	0.64	173.4	124.4	2.71		−215	428	+213	+0.30
3					90				
	0.33	175.3	124.4	5.31		+215	1664	+1879	+2.63
4					45				
	0.26	200.0	132.0	7.69		+610	3298	+3908	+5.46
5					0				
	0.31	197.0	136.9	6.35		+610	2160	+2770	+3.87
6					45				
	0.37	186.0	118.9	5.03		+215	1546	+1761	+2.46
7					90				
	0.38	165.8	92.4	4.36		−215	1496	+1254	+1.75
8					135				
	0.38	157.9	94.5	4.16		−610	1329	+719	+1.01

The stop watch time for the whole event was 1·58 seconds which compares with 1·63 seconds from the film. This gives a working check on the accuracy of the results of Table 4.

From the results, positions 4–5 produce the greater tension in the arms and illustrate why strong arms and hands are needed for this event. Position 8 represents the critical phase of the movement for it is here that the performer pulls himself in towards the bar to reduce his moment of inertia and thus increase his velocity (see Chapter 4) to help his body regain its starting position.

Direct measurement

Simple direct measurements of the forces involved in human movement can be made with spring balances or dynamometers as they are sometimes called. For example, by attaching one end of a nylon cord to a swimmer's back and the other end to a fixed spring balance, an estimate of the average propulsive force can be made by asking the swimmer to swim against the cord. If the maximum force is required a slider mounted round the spring balance will record this particular value.

It is obvious that the use of dynamometers to measure forces in this way is limited, and they certainly cannot be easily used to gain force-time information. Force platforms become essential for this sort of measurement.

Force platforms

A force platform is, of course, just a plane surface whose displacement due to a force acting on it can be measured so as to give information about that force. Bathroom scales are a very simple form of a force platform, but they are only useful for measuring static forces such as the weight of a person. However, early force platforms were mechanical in their operation, and it is only recently that strain gauges and transducers have been introduced to measure the platform displacements. This has produced improvement in the accuracy of recordings, for the platforms have been able to be made more rigid because strain gauges and transducers can measure very much smaller displacements. Four types of force plat-

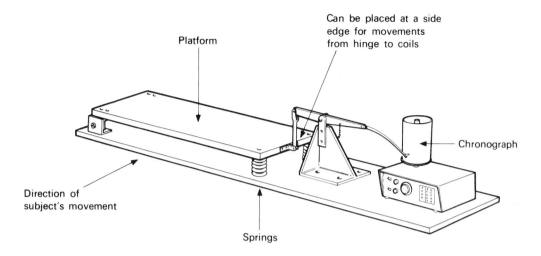

Fig. 2.15 *A mechanical force platform.*

form can be distinguished, and we shall have a brief look at each.

MECHANICAL PLATFORMS Fig. 2.15 shows a typical platform of the mechanical variety. It consists of a 60 cm square hinged platform supported at its free edge by four strong coil springs. Under the impact of a subject jumping from the platform or running across it, the platform compresses the springs and this movement is relayed by the lever system to a chronograph. The level system also magnifies the movement by a factor of around 5 so that the maximum movement of the platform, which should not exceed $\frac{1}{2}$ cm otherwise it begins to disturb a subject's performance, will produce a movement in the chronograph pen of about 2 cm. The subject's movement should always be at right angles to the free edge supported by the coils as the platform's movement then affects his performance least. Care must be taken to ensure the lever joints are not loose so that they lose any of the platform's movement. One obvious defect is that such a platform will only measure the vertical component of the force. Calibration is carried out by statically loading weights of known value centrally on the platform and noting the movement of the chronograph pen so that a suitable scale can be drawn on the chronograph. However, this scale will only give accurate values of the impact force if that force is delivered to the platform centrally. For off-centre impacts, the scale must be multiplied by the ratio x/y where x is the distance from the hinge of the platform to the static loading point, and y is the distance from the platform's hinge to the point of impact. This point can be ascertained in various ways. One way is to matt black the platform's surface and then powder the soles of the subject's shoes with french chalk. This is obviously another defect of this type of platform.

It is possible in events like the sprint start to measure the horizontal component of a performer's thrust using a mechanical platform. This was fairly successfully achieved by Henry,[4] who used starting blocks made of a wooden face plate mounted on a metal support which could slide over a base of roller bearings. Fig. 2.16 shows a platform based on Henry's design. As each block moves back under pressure, it presses against the strong spring coils. The horizontal movement is then measured by a chronograph whose pen is operated by a rack and pinion system which relays and magnifies the horizontal movements of the blocks four-fold. Some typical start figures gained by this sort of platform are as follows. A rear-block movement of 1 cm producing a pen-movement of 4 cm under the action of a 780 N horizontal thrust and a front-block

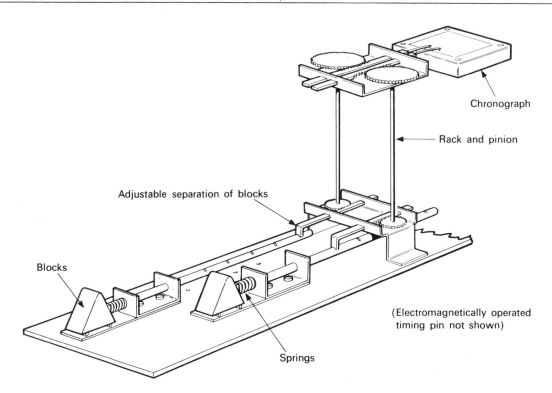

FIG. 2.16 *Mechanically operated force platform starting blocks.*

movement of ½ cm resulting in a pen movement of 2 cm under the action of 400 N horizontal thrust.

By having a second pen actuated by an electromagnet at $\frac{1}{50}$ second intervals it is possible to automatically mark the chronograph record at 0·02 second intervals, so that the force record can become time related.

CANTILEVER STRAIN-GAUGE PLATFORMS Probably the best known examples in this country of this type of platform are those of Whitney and Payne. Whitney's platform is housed in the Biomechanics Laboratory at the National Institute for Medical Research Laboratories at Holly Hill and with this platform Whitney[5] has investigated the strength of the lifting action in man as well as many other facets of human force patterns. Payne's platform was built originally by D. M. Williams[6] as an adaptation of Whitney's platform and has been utilised by Payne to investigate such activities as sprinting, shot-putting, jumping and weight lift-

FIG. 2.17 *A cantilever force platform.*

ing.[7] Fig. 2.17 shows the actual platform which measures 1·07 metres square and weighs 330 N. It is built of two layers of corrugated aluminium sheet sandwiched between three layers of flat aluminium, all riveted together to make a platform 5 cm thick.

The platform is suspended horizontally and vertically from 12 cantilevers by 1·6 mm wire cables as shown in the plan view. Fig. 2.18 gives details of the construction of these cantilevers, which under the action of a force being applied to the platform, undergo deformation. They are

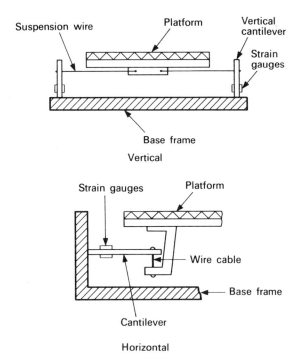

FIG. 2.18 *Vertical and horizontal cantilever supports of Payne's platform.*

under consideration the response has been found to be linear for the range of force for which it has been used.

In the initial trials, recordings were made indoors and the Wheatstone Bridge signals were fed into a slow responding pen-recorder to produce a trace. Later trials were conducted outdoors in a more natural environment and the accuracy of the recordings improved by using a fast responding ultra-violet recorder.

It is normal to link force-traces to a cine record of the activity under investigation and with this platform, Payne arranged this by placing a continuous-movement clock in the field of view of the camera, which sends pulses at known intervals to the ultra-violet recorder, which in turn records them on the force-trace.

Fig. 2.19 shows the results Payne obtained for a jump with limited arm action (Jump No 2) and with natural arm action (Jump No 1). As Payne comments "the use of the arms merely superimposes one extra late peak on to the general leg and body curve",[7] a fact hitherto undetected by indirect analysis.

so arranged however, that the vertical, the horizontal forward and the horizontal transverse thrusts produce deformations in separate groups of cantilevers, with a minimum amount of "crosstalk" between the groups, so that these components can be measured independently.

The deformation in the cantilevers is measured by strain gauges strapped to the front and rear of each cantilever and linked to a Wheatstone Bridge. Strain gauges are merely electrical conductors whose resistance changes with the amount of stretching or compression they undergo. They can be made from foil or wire and have different configurations according to the function they are expected to perform. A Wheatstone Bridge on the other hand is an electrical device which measures the "strain" of a strain gauge in terms of an electrical signal which can be used to drive a pen or ultra-violet recorder. Thus by calibrating the recorder by static loading of the platform, the vertical and the horizontal forward and transverse thrusts produced by a subject can all be measured. For the platform

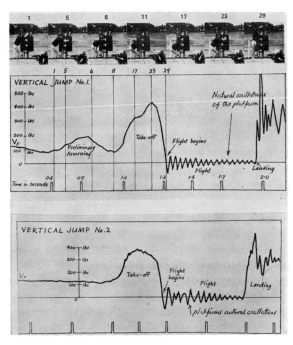

FIG. 2.19 *Force-time trace of a vertical jump (a) with arm action and (b) with no arm action.*

RIGID STRAIN GAUGE PLATFORMS Early work on this type of platform was carried out by Cunningham and Brown[8]. More recently Payne and Blader,[9,10] have built a rigid 76 cm platform for the investigation of general events such as weight lifting, shot-putting and jumping as well as a pair of force plates in the form of starting blocks for the specific investigation of sprint-starting.

(a) *The 76 cm Square Platform of Payne and Blader*

Payne and Blader developed their platform to cope with the wide variation of foot placing experienced in analysing running and jumping events. Their platform is made of 15 cm deep aluminium honeycomb of the type used in aircraft design, which has the advantage of being light (Payne and Blader's platform weighs only 300 N) as well as having a very high natural frequency (approx. 400 Hz). It is essential that the natural frequency of any force platform should be greater than that of the forces being measured so that the two can be easily distinguished. As a working guide, human movement usually produces frequencies of the order of 30 Hz so that a good platform should have a natural frequency in excess of 200 Hz. Short brass tribular posts are used to support Payne and Blader's platform because they are not prone to corrosion, can be made with uniform cross-sections, possess high natural frequencies and allow versatile placing of the strain gauges for temperature compensation and maximum decoupling of the forces being measured. That is, the force being measured along one axis produces significant readings in one group of gauges but not in the other groups mounted to measure forces along other axes.

The supporting base of the platform is made from 1 cm thick mild steel and is fitted with three feet so that uneven surfaces do not present a levelling problem. The positioning of the gauges is rather complex, but the signals fed to the ultra-violet pen recorder (used in preference to a mechanical pen recorder to minimise trace distortions) record the horizontal and vertical force being developed as well as the couple about the vertical axis.

(b) *Payne and Blader's Force Plates*

In contrast to the rigid platforms, Payne and Blader's force plates are supported on semi-circular elements to which the strain gauges are fitted. These gauges are fed by a 3 Hz a.c. and are fitted into Wheatstone Bridge circuits, whose unbalanced signals are amplified and fed into an ultra-violet pen recorder that gives a normal and tangential force trace for each foot together with the couple about the horizontal axis.

Connected to other galvanometers in the ultra-violet recorder and giving appropriate signals on the paper are a microphone to pick up the sound of the starter's gun, three photo-electric timing beams to record the athlete's progress down the track and a large visual clock to synchronise cine

FIG. 2.20 *Payne and Blader's force plates for investigating the sprint start.*

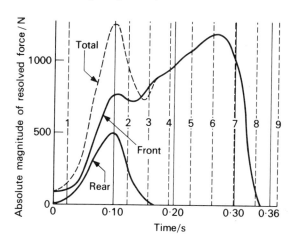

FIG. 2.21 *Calculated values of the magnitude of the resolved forces for the front foot, rear foot and both feet.*

film and paper recording. This is an essential feature as force traces alone are not much use in analysing performances.

A typical trace obtained with the force plates is shown in fig. 2.21 which shows the calculated magnitude of the resultant forces produced by the forward foot, the rear foot and both feet for this particular sprint start. Note how both feet exert a force at the same time.

The angle of action of these forces can also be calculated and when this is done it is possible to draw in the actual force acting on synchronised frames of the cine record. Fig. 2.22 shows the result of doing this for the sprint start considered earlier.

Note how:

(1) the force vector lies first in front of the centre of gravity and then behind, giving the sprinter a helpful backward rotation followed by a small unwelcomed forward rotation, that he has to counterbalance by his arm action.

(2) the angle of lean decreases as the acceleration decreases as the average horizontal force becomes less and less.

From a study of some 150 athletes, Payne and Blader also concluded that in the sprint start both feet exert a force at the same instant (see fig. 2.21) and that better starts were characterised by a strong rear leg action, which is a case of *not* "putting your best foot forward"!

Both the force platform and force plates give a linear calibration which is another important design feature as a non-linear response platform makes trace interpretation very difficult. Again, as it is necessary for environmental conditions to be as natural as possible, so that activities normally conducted outside should not be carried out in a small room, the strain gauges of both the platform and the plates are waterproofed so that the systems are not influenced by dampness. This allows both systems to be used outside and in addition the platform to be sunk into the ground.

TRANSDUCER PLATFORMS   One of the difficulties of the platforms discussed so far has been that they can only be used in one place and can only record a limited part of a subject's movement, albeit the important part.

Equipment developed by the Royal Aircraft Establishment and marketed by Lloyd Instruments Ltd[11] overcomes this difficulty by fitting the force platform into the soles of the subject's shoes.

Known as the Stanmore Sandals, the soles of the sandals are fitted with load measuring transducers, formed by sandwiching flexible metal filigree sheets between layers of rubber sponge. As force is applied to the soles of the sandals the rubber sponge is compressed and the transducer's

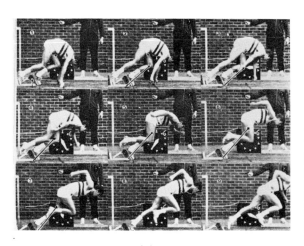

FIG. 2.22 *Force vectors determined from force plate analysis illustrated by fig. 2.20 drawn on the synchronised cine frame of fig. 2.21.*

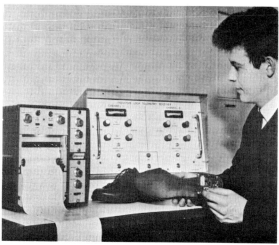

FIG. 2.23 *The Stanmore Sandals*

electrical capacitance is altered. The change in capacitance affects the frequency of the transmitting circuit of which it forms a part and a nearby loop aerial picks up these changes and feeds them into an ultra-violet pen recorder. The strength of the signal fed into the recorder is directly proportional to the force being applied to the soles of the sandals at any given time. Fig. 2.23 shows the uncovered sole of one of these sandals together with the receiving and recording equipment.

Each sandal's transmitter is limited to a particular frequency transmission band so that the receiving equipment can differentiate the force-trace being sent out from each sandal. The frequency bands are fairly low because the transmission takes place near the ground and because it must conform with G.P.O. regulations. Calibration shows that the changes in the signal received by the recorder are linearly related to the force changes experienced by the sandals and are virtually independent of the area of sole in contact with the floor. The aerial loop is effective if placed round a 18 m × 3 m area near to the floor but this area must be free from pieces of "buried" metal.

Underwater weighing devices

So far we have discussed techniques designed to measure ground thrust forces. To complete this chapter, let us look at devices designed to measure weight-in-water forces or more simply body densities. Such measurements are important medically as well as physically because a person's buoyancy not only tells us how well they can float, but also how obese they are or what is the value of their lung's vital capacity.

Direct methods for measuring density can be separated into two classes—those which measure the amount of water displaced and hence record a direct measure of body volume, and those which measure the apparent loss of weight of a subject immersed in water and therefore record the value of the water's upthrust which is an indirect measure of body volume. Strictly it is this latter class that are underwater weighing devices and these can be further subdivided into those devices that use a suspension technique and those which use a platform technique.

Water Displacement Methods

At an elementary level, people can use this method to find their body density by weighing themselves on a pair of bathroom scales and then noting how much the water in their bath rises when they get into it and fully exhale or fully inhale and submerge themselves. Most baths can be treated as rectangular so that their body volume will be equal to the product of the rise in the water level, the length and the breadth of the bath. At a slightly more advanced level, a cylindrical tank, about 2 m high and 0·7 m diameter, can be used.

Body density values will of course be given by the equation:

$$d_b = \frac{\text{Subject's weight}}{\text{Subject's body volume}} = \frac{W_b}{V_b}$$

However, if we are interested in the water's upthrust, U, this will be given by Archimedes' Principle, namely

$$U = V_i \times d_W$$

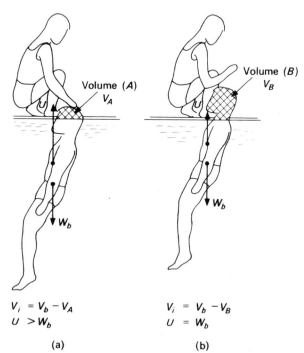

FIG. 2.24 *The buoyancy factor.*

where d_w is the water's density and V_i is the immersed volume of the subject. For a person to float unaided U must be equal to W or the ratio U/W must be unity. It is thus useful to define this ratio as a person's buoyancy factor. Then the nearer to unity this factor lies, the less upthrust the person will need to stay afloat. Should it exceed unity then U is greater than W and will therefore force the swimmer's body further out of the water, thereby reducing V_i and U until U and W are equal. Figs. 2.24(a) and (b) illustrate this point. Should of course it be less than unity then the swimmer will sink.

Returning to the buoyancy factor itself:

$$B = \frac{U}{W}$$

we can substitute from these equations for U and W_b, so that:

$$B = \frac{V_i d_w}{V_b d_b}$$

As V_i and V_b are not very different for a person floating or swimming and d_w is near enough to unity, we can write as a first approximation:

$$B = \frac{1}{d_b}.$$

This equation shows the essential relationship of body density to the buoyancy of a person and in most practical determinations of B, it is this value of B that it measured.

The value of B lies very close to unity in most cases so that very little upthrust is needed to keep afloat. Any extra upthrust that is required by a swimmer is automatically gained when he propels himself through the water from the vertical component of the swimmer's propulsive force. It can also be provided by artificial aids for swimmers unable to propel themselves, because they are beginners or non-swimmers who have fallen into the water.

Suspension Weight-in-Water Methods

Fig. 2.25 shows a simple way of determining the buoyancy factor by this technique using a spring balance and cradle. The spring balance is

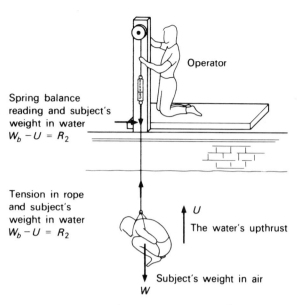

FIG. 2.25 *A simple suspension weight-in-water system.*

read with a fully exhaled position, say, and by the application of Archimedes' Principle it can be shown that:

$$\frac{\text{(Upthrust on swimmer)}}{\text{(Weight of swimmer)}} = \frac{U}{W_b}$$

$$= \frac{W_b - R_2}{W_b}$$

or

$$B \text{ exhaled} = 1 - \frac{R_2}{W_b}$$

W_b can be measured on ordinary medical scales. Care must be taken in reading the spring balance as the needle will tend to oscillate even when the subject is fully relaxed. For B inhaled the upthrust is so large and frequently greater than W, that it is difficult to get a large enough reading on the spring balance to read it with any reliability. When this is the case a sinker is attached to the cradle for a first reading (R_1) and held by the subject for a second reading (R_2). R_2 in the equation above then becomes ($R_2 - R_1$). A similar procedure should be used if the cradle is not lightweight. It is possible to improve the method if the spring balance is replaced by a cantilever with a strain gauge strapped to it, as fig. 2.26 shows.[12]

Fig. 2.26 *A cantilever underwater weighing device.*

The strain gauge is strapped to the neck of the horizontal cantilever.

Underwater weighing platforms

Suspension cradles of the type just described are prone to rather large natural oscillations although they are relatively easy and inexpensive to build. Platform equipment is less likely to suffer from this deficiency. The most commonly used type of platform equipment is based on the apparatus successfully developed by Von Dobeln[13] in 1965, which consists of a water tank, a pendulum balance and a chair in the tank fitted by two inverted U tubes to the pendulum balance's platform. With this equipment a subject immerses himself by sitting on the chair and then leaning forward till he is fully submerged. If this is done carefully natural oscillations cause no difficulty. The fear of submersion is also minimised.

A more recent extension of the platform technique has been devised by Akers and Buskirk[14] using underwater transducer cells which will give a continuous reading of the underwater weight in the same way that strain gauges will.

3 | The centre of gravity of the human body

Importance of the centre of gravity

Everything, including the human body, has a centre of gravity and in any consideration of human movement, the position of the centre of gravity of the body plays an important part. Take for example the diver of fig. 2.9 of Chapter 2. As the thrust from the board lies behind his centre of gravity it not only moves him upwards and outwards but also produces a couple $F_1 x$ that gives him the necessary forward rotation for his dive. If his positioning on the board is not correct, either this rotation will not be sufficient or, if the final thrust lies in front of his centre of gravity, it will give him backwards instead of forwards rotation.

Position of the centre of gravity of the human body

The centre of gravity of a regular body such as a rectangular block or a cylinder lies at the geometric centre of the body and is therefore easily located. For irregular bodies this is not so and their centres of gravity have to be found either by experiment or by calculation. As far as the human body is concerned, not only is it irregular but it also changes its shape as well.

For an adult male the centre of gravity lies about 2·5 cm below his naval or about 57% of his full height from the ground.[15] For the adult female the centre of gravity lies about 55% of her full height from the ground due to her heavier pelvis and lighter thorax, arms and shoulders. Growing children have more weight in the upper part of their bodies and so their centre of gravity is more like 60% of their full height. This still gives them a lower centre of gravity than adults when standing side by side.

Changes in position of centre of gravity of the human body take place with movement and tend

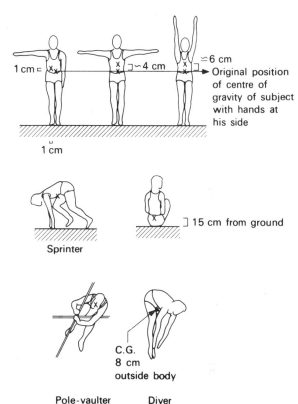

FIG. 3.1 *Relationship of the centre of gravity to the human body in different activities.*

to move in sympathy with the movement and can even lie outside the body as fig. 3.1 shows. Eating and breathing also cause small changes.

Movement through the air

In most human movement the body is either in contact with the ground or it is moving through the air. Let us first look at movement through the air. By studying the path traced out by the centre of gravity of a body, we find that it always travels on a parabolic path. Take for instance the diver shown in fig. 1.2 or the high-

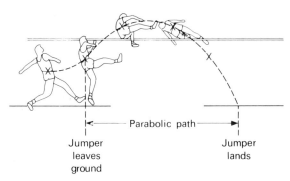

Parabolic path of Western Roll high jumper

(a)

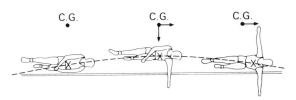

Movement of centre of gravity horizontally with respect to the head and vertically to the underside of the body at the peak of the Western Roll to gain better clearance

(b)

FIG. 3.2 *High-jumper's path through the air.*

jumper of fig. 3.2. Investigation of the curves traced out by the centres of gravity in each case will reveal they are parabolas. Notice how at the peak of his Western Roll, the high-jumper gains extra clearance by pushing his centre of gravity downwards with respect to the underside of his body by thrusting his lower arm down. In practice we can only raise our centre of gravity by about one metre in any high jump and, as a consequence, it is very important to choose a style which gives maximum clearance. In the arch straddle jump the centre of gravity passes under the bar and thus this style of jumping tends to give the greatest clearance height, all other parameters being equal.

Maximum range

In sports like the shot-put and long-jumping it is very important for a competitor to achieve maximum range for a given effort. That the curve is a parabola can be shown theoretically by referring back to the shot-putter of fig. 1.7.

It was shown that the vertical co-ordinate of the shot after t secs is given by the equation

$$y = V_y t - \tfrac{1}{2} g t^2$$

where $g = 9 \cdot 8 \, \text{m s}^{-2}$, while the horizontal co-ordinate x is given by:

$$x = V_x t$$

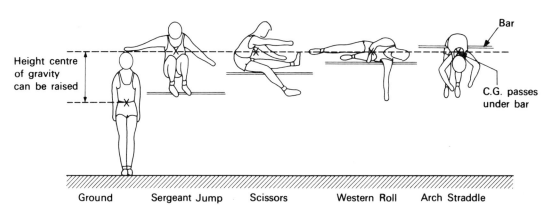

FIG. 3.3 *Bar clearance variation with style of high jump.*

Eliminating t between these two equations gives us

$$y = \frac{V_y}{V_x}x - \frac{gx^2}{2V_x^2}$$

or

$$y = x\tan\theta - \frac{g}{2V^2\cos^2\theta}x^2$$

(as the vertical component of V, is given by

$$V_y = V\sin\theta$$

and the horizontal component of V, by

$$V_x = V\cos\theta)$$

Now this equation is the equation that when plotted graphically gives a parabola. With $y = 0$ and $x = R$, it gives us:

$$R = \frac{2V^2\cos\theta\sin\theta}{g}$$

where R is the range, V is the initial velocity and θ is the angle of projection. Keeping V constant, it is possible to calculate the value of R for different values of θ and we find that R is greatest in this case when θ is 45°. But no long-jumper or shot-putter operates at this angle, so that theory and practice seem divorced. But neither the long-jump or the shot are ground to ground projections. The long-jumper's centre of gravity falls about 15 cm from start to finish while the shot's centre of gravity falls about 2 metres. This means we must put $x = R$ and $y = (-h)$ into the equation where h represents the fall of the centre of gravity below the point of projection, in order to find the range in these cases. When this is done we get

$$R = \frac{V^2\sin\theta\cos\theta + V\cos\theta\sqrt{V^2\sin^2\theta + 2gh}}{g}$$

Again keeping V constant, it is possible to calculate R for different values of θ. For a shot-putter where the shot leaves his hands approximately 2 metres above the ground, R is maximum when θ is 41°. For a long-jumper where h is 15 cms the value of θ that gives maximum R is 43°. There is still a discrepancy between these values and those used by national and international shot-putters and long-jumpers as mentioned in Chapter 1. The discrepancy occurs because the angles quoted above have been derived assuming that the initial velocity V is constant whatever the angle of projection. This assumption is not true of either shot-putters or long-jumpers. The initial velocity will vary with θ, and if we make some assumptions how it does vary with θ and substitute for V by such an equation and start all over again, the discrepancy between theory and practice starts to disappear.[16] This serves as a good illustration that a little knowledge can be a dangerous thing. A coach suggesting a jumping or putting angle of 45° would certainly not get maximum results for his protegées.

Movement on the ground

Movement on the ground tends to be concerned with the human body being in a unstable or stable state, where there are two basic types of stability. The first—static stability—is exemplified by a gymnast in a headstand, and the other—dynamic stability—by a gymnast walking a tightrope—while instability is exemplified by a gymnast falling into a forward somersault.

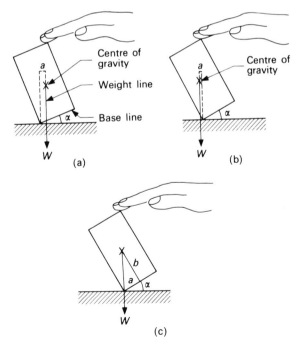

Balancing a block of wood

FIG. 3.4 *Balancing a block of wood.*

Static stability

Consider the block of wood shown in fig. 3.4(a). On release it will fall back to its original position because the couple "$W.a$" acts in a clockwise direction. This will happen until it is released from the position shown in fig. 3.4(b) when the couple "$W.a$" begins to act in an anti-clockwise direction causing the block to fall over on to the table. Up to the point where the block returns to its original position it is said to be statically stable. This will occur as long as the weight line from the centre of gravity passes through the base line. As soon as it passes outside the base line the block will fall over.

Therefore, no body with a weight line from its centre of gravity that lies outside its base can be stable. Returning to the block of wood, its stability is increased the larger the tilting angle α becomes before the block topples over on release. The toppling angle α_m illustrated by fig. 3.4(c) is given by

$$\alpha_m = \tan^{-1}\frac{a}{b} = \tan^{-1}\frac{\text{base}}{\text{height}}$$

A large α_m is given by a block that has a large base and short height. This is true generally for all regular or irregular bodies, that good static stability is given when the body concerned has a large base and its centre of gravity is as near to the base as possible. A gymnast doing a headstand as opposed to a handstand illustrates this principle well. Also, all other things being equal, a heavier body is more stable than a light one as the couple needed to tilt the body by the same amount is necessarily larger, and with irregular bodies stability is increased by the centre of gravity lying as far away from the tilting edge as possible.

In human postures enlarging the base area frequently lowers the centre of gravity as fig. 3.5 shows. Also when involved in activities such as weight-lifting the body adjusts its posture to bring the weight line of itself and the barbells back into the base area. Several examples of this are shown in fig. 3.6 which all involve a lean of some kind.

As we have introduced the idea of a combined centre of gravity in fig. 3.6 this is probably the best point to indicate how a combined centre of

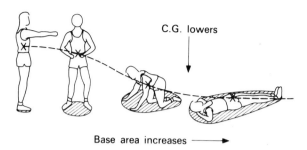

FIG. 3.5 *Stability increase with enlarging of base area and lowering of c.g.*

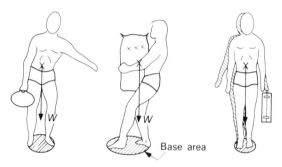

(a) Lifting postures - note lean of subjects

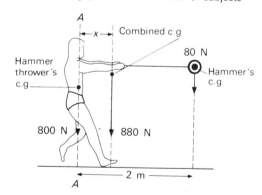

(b) Hammer thrower's combined centre of gravity

FIG. 3.6 *Balancing loads.*

gravity can be found when the centres of gravity of the components are known.

It follows from Chapter 2 that the moments produced by the separate components must equal the moments produced by the combination about any point that is chosen. So for the hammer thrower of fig. 3.6.

THE CENTRE OF GRAVITY OF THE HUMAN BODY

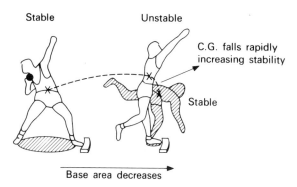

(a) Shot-putter's stable - unstable - stable movement

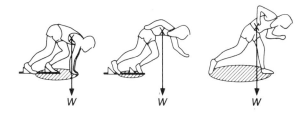

(b) Sprinter's stable - unstable - stable movement

FIG. 3.7 *Stable to Unstable States.*

Separate Moments about A = Moment of the combination about A

$$(80x2) + (800x0) = 880x$$

$$x = \frac{160}{880} \text{ metre}$$

$$= 18 \text{ cm}$$

Many of the movements that we make require us to go from a stable state to an unstable one and then back again to a stable state. Examples of this are numerous but fig. 3.7 illustrates two examples of this.

Dynamic stability

Dynamic stability, as exemplified by a tightrope walker, is best discussed with reference to fig. 3.8. Standing on a bench the smallest rod is statically the most stable. However, if we try and balance the rods on the end of one of our fingers, it is easiest to balance the long rod. So the long rod is the most stable rod dynamically. This is because on the end of our fingers a rod falls over

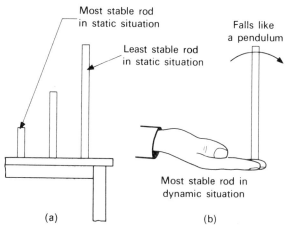

FIG. 3.8 *Dynamic stability.*

like a pendulum and the longer the rod, the longer it takes to fall, giving the finger more time to put the base of the rod back under the rod's centre of gravity. How the tightroper achieves this will be discussed in Chapter 4, but basically he moves his body to keep his centre of gravity immediately over the tightrope rather than by moving his base.

Centre of gravity determination

There are basically two approaches to centre of gravity determinations; the first uses the human body itself while the second uses suitable models as a substitute.

Centre of gravity board method

First let us look at what can be called a modified centre of gravity board technique because it

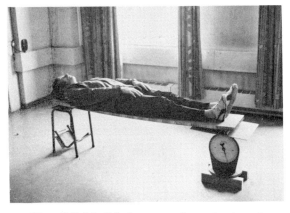

FIG. 3.9 *Modified centre of gravity board.*

restricts the determination to the vertical position and movement of the centre of gravity. The apparatus needed is illustrated in fig. 3.9. It consists of a 2·13 m × 0·3 m × 2·5 cm piece of F-Web board[17] resting on a fulcrum placed on a set of parcel scales at one end and on an angle-iron hinge at the other end. By applying the principle of movements about the hinge to a subject lying on the board as shown in Figure 3.10(a)

$$W \cdot x_1 + w \cdot x = E_1 \cdot d$$

where W is the subject's weight (previously measured on the parcel scales) x_1 is the distance from the hinge and the subject's feet to his centre of gravity, w is the weight of the board, x is the distance from the hinge to the board's centre of gravity, E_1 is the parcel scales' reading and d is the distance between the hinge and fulcrum. If the parcel scales are now read E_0 without the subject lying on the board, the principle of moments applied again about the hinge, yields

$$w \cdot x = E_0 \cdot d$$

Eliminating "wx" from the equations gives

$$x_1 = \frac{(E_1 - E_0)d}{W}$$

As all the quantities on the right-hand side of the equation are known, x_1 can be calculated and this represents the vertical height of the subject's centre of gravity above his feet.

If the interest is centred more on the vertical movement of the centre of gravity such as that involved in moving from the position shown in fig. 3.10(a) to that of fig. 3.10(b), the scales are read E_1 for the first position, and E_2 for the second position with the hands above the head. With the hands above the head x_1 becomes x_2 and the scales read E_2, so that:

$$W \cdot x_2 + w \cdot x = E_2 \cdot d$$

on taking moments about the hinge yet again. By subtracting the two equations we get:

$$\Delta x = x_2 - x_1 = \frac{(E_2 - E_1)d}{W}$$

where Δx represents the vertical shift of the subject's centre of gravity. E_0 is not involved in this equation and therefore is not needed. Typically a man's centre of gravity will be raised 7 cm by raising his arms above his head.

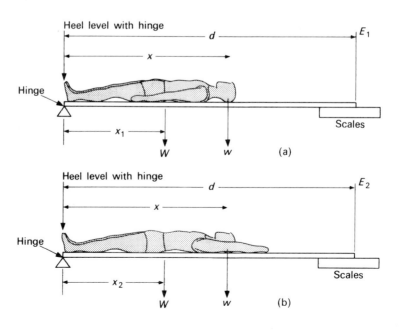

FIG. 3.10 *Modified centre of gravity board with subject raising his arms.*

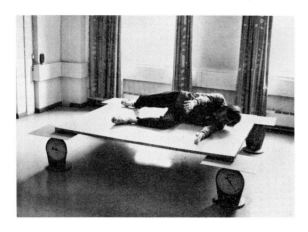

FIG. 3.11 *Full centre of gravity board.*

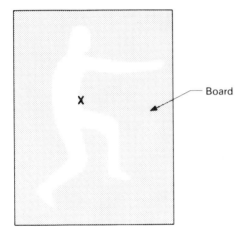

FIG. 3.12 *Position of the centre of gravity for subject in position shown in fig. 3.11.*

Only the vertical position and movement of the centre of gravity can be investigated by the modified board, although for symmetrical positions like those of fig. 3.10 a large error is not introduced if it is assumed that the vertical position of the centre of gravity lies in the plane that bisects the body into two equal parts.

A two-dimensional fix on the centre of gravity demands the use of a board like that shown in fig. 3.11. It consists of a rectangular piece of rigid board with a set of parcel scales at each corner. A small stop fixed to the board at each corner allows a good central contact between board and scales to be made. With the subject in position as shown in fig. 3.11 the position of his centre of gravity can be found by recording the parcel scale readings with him on and off the board and by measuring the length and breadth of the board. By taking moments about different edges of the board, the position of the centre of gravity in two dimensions can be calculated. Fig. 3.12 shows where these results place the centre of gravity for the position and subject considered.

If a centre of gravity trace is required for a particular activity, a film must be taken from the side or more rarely the front of the movement. When the film is developed, selected frames are projected on to the centre of gravity board, which is held in a vertical position, so that life size images of the subject are obtained. A coloured chalk outline is then drawn round a given image of the subject and the board is placed horizontally on the parcel scales. The subject then gets into the position marked on the board and the necessary readings and calculations that have already been outlined, are made so that the position of his centre of gravity can be found and marked on the outline on the board. If a selection of frames is dealt with in this way a centre of gravity trace can be obtained.

The method of projection of images can be the same as for velocity analysis but the distance between projector and board with this method is fairly large and the intensity of the image rather low. An alternative method is to project an image on to an overhead projector transparency and then to use this transparency on an overhead projector to produce the necessary life-size image.

If the movement cannot be confined to the board with the board and projector remaining in fixed relative positions to one another, a grid system has to be introduced. One way of doing this is to film a given activity against a grid background, such as a black wall painted with a white grid. An alternative is to use the overhead projector method. The frames of the film are thrown on to a vertical ground glass screen with a grid already marked on it. The selected frames are then projected on to this screen and the subject outlines are marked on to it. Overhead transparencies can then be made with both the subject outline and suitable grid reference lines drawn in. These in turn can be projected on to the centre of gravity board via the overhead projector.

When the position of the centre of gravity is subsequently marked on to the board its grid reference can be read and transferred to the ground glass grid. As it is rather wasteful to make the master trace on the glass screen, the screen can be covered with tracing paper and the master trace can be drawn on this instead. The projection of images is of course a much larger source of error than the inconsistency of readings mentioned earlier.

One other source of error is that the subject can never reproduce the position he was in during the actual movement when he lies down on the outline. In fact in movements like the backward somersault, the subject finds difficulty in statically holding some of the positions that he went through dynamically and may well have to be strapped into several positions.

Yet another source of error which is probably not large, but is unknown, is that due to the redistribution of body mass between dynamically moving through a position and holding it statically.

As a consequence of these errors, the centre of gravity board is not so highly accurate as may at first be thought. Therefore, as it is a very time consuming method, the manikin methods about to be described have much to recommend them on this front.

Manikin methods

A manikin is a jointed model of a man, often in two dimensions only. Manikins usually represent either a front view or a side view, and as no human being has exactly the same body characteristics, there is no general manikin that can be made and used in all centre of gravity analyses with complete accuracy. However, researchers have compiled values of relative limb mass and relative positions of centres of gravity, that have been "refined" to give more and more consistent results until it is possible to make a standard manikin that can be used with a fair degree of accuracy for centre of gravity analyses involving mesomorphic subjects that lie in the mass range of 60 to 80 kg.

Figs. 3.13 and 3.14 show the templates that can be used to make two such manikins and these are based on those suggested by W. T. Dempster[18], M. Williams and H. Lissner[19] and Dreyfuss[20]. The figures for the position of the centre of gravity and limb mass for these are based on the researches of W. T. Dempster[18], H. T. Hertzberg[21] and the Bioastronautics Data Book of the National Aeronautics and Space Administration, taking into consideration the refinements suggested by Dr R. J. Whitney[22] for the figures to give consistent results. The manikins can be made in either cardboard or plywood.

The limitations of such manikins are governed by the generalised results on which they are based. It is possible to improve accuracy slightly by using a more individualised manikin, i.e. one based on a particular subject.

To make an individual manikin, the two sets of figures displayed on the manikins of Figs. 3.13 and 3.14 have to be collected for the particular subject under investigation. These two sets of figures are the centres of gravity of the segments, and the weights of the segments. For many purposes the segmentation of Table 5 is sufficient—

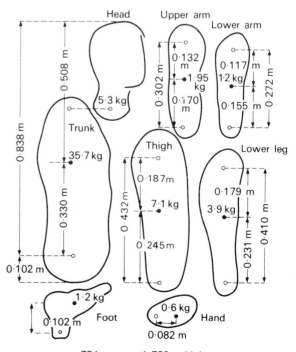

FIG. 3.13 *Side elevation template for an average man.*

THE CENTRE OF GRAVITY OF THE HUMAN BODY

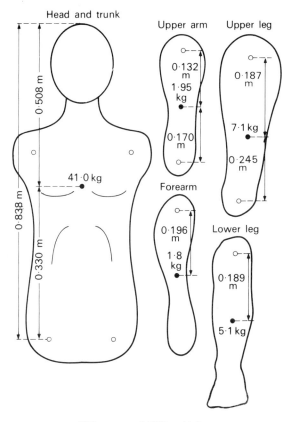

73 kg man : 1·789 m high

FIG. 3.14 *Front view template of an average man.*

Table 5

Individual manikin segmentation

Segment	Location
Head and trunk	from tip of head to hip joint
Upper arm	from tip of shoulder to elbow joint
Forearm and hand	from elbow joint to tip of first finger
Thigh	from hip joint to knee joint
Lower leg and foot	knee joint to heel (foot at right angles to leg)

this is slightly modified to the segmentation of the standard manikins in that the lower leg and foot, the forearm and hand, and the head and trunk are treated as single segments. Greater segmentation is not permitted by the procedures now to be outlined.

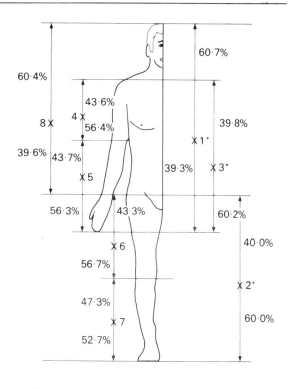

Key
1. C.G. of trunk, head and arms at side
2. C.G. of whole leg
3. C.G. of whole arm
4. C.G. of upper arm
5. C.G. of forearm and hand
6. C.G. of thigh
7. C.G. of lower leg and foot
8. C.G. of head and trunk

FIG. 3.15 *Percentage position of centres of gravity of body segments.*

(a) THE CENTRES OF GRAVITY OF THE SEGMENTS
These cannot be measured directly unless the body is dissected and a gravity board used to find the centres. However, the research workers mentioned earlier have compiled tables which give the positions of the centres of gravity of each segment. These positions have been expressed in percentage terms and average values have been gained for different body weight groups. Fig. 3.15 gives the results for a 73 kg man. Provided a subject lies within the range 0–80 kg these figures do not vary much and can be used for the whole range without great inaccuracy occurring. The starred percentages which appear on the right hand side of fig. 3.15 will be referred to later.

Thus to find the position of a segment's centre of gravity for an individual, it is only necessary to measure the length of the segment concerned and then using the percentage quoted for that segment in fig. 3.15 to calculate the distance to the centre of gravity.

(b) THE SEGMENT MASSES These can be found by using similar percentage figures but referring to mass rather than centres of gravity. However, the net result is really to produce a manikin very similar to the standard manikin already described. A study of these figures also shows a greater variation in percentage terms than they do for the position of the centres of gravity, and thus they introduce a larger error. It is therefore better to determine these masses experimentally, using a modified Centre of Gravity Board.

To obtain the necessary segment weights position the subject as shown in fig. 3.16(a), referred to as position A_1. Record the reading on the set of scales and without moving the subject's trunk position, help him into the position shown in fig. 3.16(b) referred to as position A_2 and note the new reading on the scales.

In position A_1 taking moments about the hinge gives:

$$W_1 x_1 + W_2 x_2 + W x = E_1 d$$

where the length of the board is d. For position A_2:

$$W_1 x_1^1 + W_2 x_2 + W x = E_2 d$$

where W_1 = weight of the two thighs and lower legs

W_2 = weight of the rest of the body

W = weight of the centre of gravity board

x_1 = distance from hinge to the centre of gravity of the thigh and lower leg in position A_1

x_1^1 = same distance in position A_2

x_2 = distance from hinge to the centre of gravity of the rest of the body in position A_1 and A_2

x = distance from hinge fulcrum to the centre of gravity of the board

d = distance from hinge fulcrum to the scales' fulcrum

E_1 = scales' reading in position A_1

E_2 = scales' reading in position A_2

Subtracting the two equations gained above then yields the segments weight W_1 as:

$$W_1 = \frac{(E_2 - E_1)d}{(x_1^1 - x_1)}$$

As E_1 and E_2 and d are known, W_1 can be calculated if $(x_1^1 - x_1)$ is known. The distance $(x_1^1 - x_1)$ represents the horizontal distance the centre of gravity of the thigh and lower leg has moved and this is the same distance as the distance between the centre of gravity of the thigh and lower leg and the hip joint. This distance can be calculated from the relevant starred percentage shown in fig. 3.15 (i.e. 2*). For example, for a 2-metre-high man, this distance works out to 39 cm. It should be noted that W_1 represents

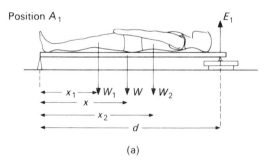

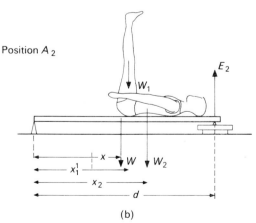

FIG. 3.16 *Body segment weights.*

the weight of both legs and thus the weight of one is given by $W_1/2$. If one limb is moved then W_1 represents just one limb. The difference in the weights between left and right limb does not usually warrant this procedure, which necessitates carrying the procedures through first for the left limbs and then for the right limbs. It also demands very sensitive scales because the changes in E_1 and E_2 are a factor two times smaller.

If the sequence discussed above is followed through for the four sequences shown in fig. 3.17, W_1 will represent in turn

(a) twice the weight of the lower leg and foot,
(b) twice the weight of the whole arm,
(c) twice the weight of the forearm and hand,
(d) the weight of the trunk, head and two arms.

The judicious use of straps as illustrated in the last sequence of fig. 3.17 can help to keep the rest of the body in the same position between the two scale readings E_1 and E_2, as well as in the latter case, the part of the body being moved.

In order to calculate W_1 for these segments it is necessary to know the distances $(x_1^1 - x_1)$ for each of the sequences. These distances represent in turn:

(a) the distance from the knee joint to the centre of gravity of the lower leg and foot (7);
(b) the distance from the shoulder to the centre of gravity of the whole arm (3*);
(c) the distance from the elbow joint to the centre of gravity of the forearm and hand (5);
(d) the distance from the hip joint to the centre of gravity of the trunk, head and arms with the arms held at the side (1*).

They can be calculated by using the starred percentages of fig. 3.15, as indicated in parenthesis.

With these distances the weight of the segments of the body mentioned earlier can be found and, by subtraction, the weight of the head and trunk, the upper arm, the forearm and hand, the thigh, the lower leg and foot can be determined. Table 6 gives the results gained for a 2 m tall subject.

A useful check on accuracy is to compare the total weight of all the segments with the actual weight of the subject (Table 6). It is possible to make the individual manikin model from either

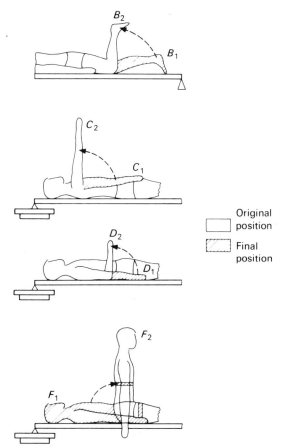

Positions $B_1 - B_2 : C_1 - C_2 : D_1 - D_2$ etc. which yield the segment weights of the lower leg and foot, the arm and hand, the forearm and hand and the head, trunk and arm.

FIG. 3.17 *Centre of gravity board sequences for determining body segment weights.*

Table 6
Final segment weights for a 2 metre 79·54 kg wt man

Segment	kg wt
Mass of head and trunk	42·73
Mass of upper arm	2·18
Mass of forearm and hand	1·91
Mass of thigh	9·75
Mass of lower leg and foot	5·14
Total weight	80·69

a front or side position. A slide is needed of the subject standing to attention taken from the side or front as the case may be. This slide is then projected on to a screen of white cardboard or plywood to a suitable size. A useful scale is 1 to 6 cm. Thus for the subject mentioned earlier who is 2 metres tall, the size of the projected image will be 33 cm high. An outline of the image is made and the individual segments needed to make the manikin are taken from this outline. There are two ways the centre of gravity can be marked on the segments. It can be made directly by marking them on the subject in black ink before the slide is taken; or it can be done indirectly after the segments have been cut out by scaling down the distances and marking the centres of gravity on lines drawn between the joints of each segment. The segments are then joined together as for the standard manikin.

Provided the modified gravity board readings are carefully carried out, an individualised manikin is more accurate in use than a general one, but obviously takes more time to make.

In order to make a centre of gravity trace using a manikin, a film of the subject in action has to be taken. As it is necessary to have an outline of the subject to make the manikin as well as to get the projected images of the subject to the exact size of the manikin, it is useful, but not essential, to film the subject standing to attention before he moves into the activity for which the centre of gravity trace is needed. This part of the film can then be used for this purpose.

When the manikin has been made and a tracing screen has been fixed at a distance from the projector that gives images the same size as the manikin, selected frames can be projected and outlines drawn in the same way as for the centre of gravity board.

The determination of the position of the centre of gravity for a given outline can be calculated by applying the principle of moments to the composite manikin, using the appropriate data, first in the x direction and then the y direction. When the x and y values have been found they can be used to mark the centres of gravity on each outline and the joining together of these points will yield a centre of gravity trace.

An alternative method that does not involve such irksome computations demands a slightly different type of manikin.

The manikin used in this method needs to be made in a slightly different way in that the segments must be proportional in weight to their real counterparts. This can be achieved by small metal discs being fixed to the manikin's segments at the centre of gravity mark. When the resulting manikin is suspended as shown in fig. 3.18—freely enough to swing back to its initial position if moved to one side—its centre of gravity must be somewhere along the plumbline. For when pulled to one side, the weight of the manikin which acts from the manikin's centre of gravity will immediately produce a restoring couple "Wa" that will rotate the manikin back to its rest position. This couple only becomes zero when "a" becomes zero and this only occurs when the manikin's centre of gravity lies directly beneath the point of suspension, i.e. somewhere on the plumbline. If the manikin is now suspended from another point, its centre of gravity must lie on the new line marked out by the plumbline. If these two lines are drawn on to the manikin, the point where they cross must be the manikin's centre of gravity

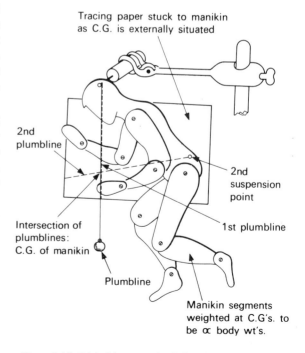

FIG. 3.18 *Plumbline method for determining c.g.*

as it is the only point that is common to both lines. A check on accuracy can be made by suspending the manikin from a third point. The line now traced out by the plumbline should also pass through the intersection of the other two lines as well.

If the outlines of the subject have been made as in the previous section, the manikin can be fitted into the outlines and when the centre of gravity has been determined for one outline by the method just described, the manikin can be placed back over the outline. A pin pushed through the intersection of the two lines made by the plumbline which marks the position of the centre of gravity, will mark the relevant centre of gravity on to the outline. When this has been done for all outlines, all the pin marks can be joined together to give the required centre of gravity trace.

For some positions where the centre of gravity lies outside the manikin a square of tracing paper stuck to the manikin can overcome the difficulty of the plumblines intersecting off the manikin.

Using a weighted manikin, K. Lane[23] has carried out a comparative study between the centre of gravity traces obtained with the manikin and a centre of gravity board. Fig. 3.19 shows the results he obtained for one subject doing a front somersault. Note the parabolic nature of the performer's path. This is confirmation that all bodies moving through the air follow a parabolic path whether they be divers doing a pike dive or gymnasts executing a forward somersault (providing air resistance is negligible).

Diagnostic use of centre of gravity traces

One of the most successful analysis techniques in diagnosing weak spots in gymnastic movement has been the analysis of the path of the centre

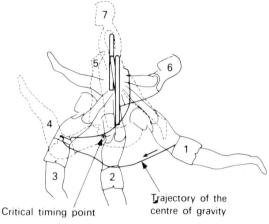

(a) Good movement

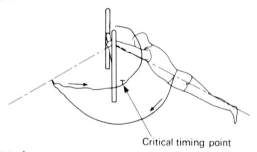

(b) Average movement

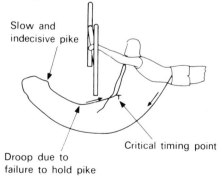

(c) Weak movement

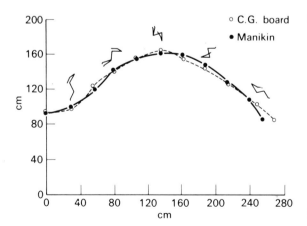

FIG. 3.19 *Comparison of a c.g. trace produced by a subject with the c.g. trace produced by a manikin of the subject.*

FIG. 3.20 *Centre of gravity analysis of a gymnastic up-start movement for a good, average and weak performance of the movement.*

of gravity. By comparing the traces of a good performer of a particular movement with those of a learner having difficulties, it is possible to pinpoint specific as well as more general weaknesses. This is of obvious help to the coach who hitherto has had to rely on his own visual analysis of events which very often last no more than a few seconds. Fig. 3.20 shows the results of an analysis carried out by R. N. Harris[24] of the gymnastic up-start for a good, average and poor performer. For the good performer it can be clearly seen that after the legs were piked and the body was in its backswing the straightening of the body was timed to coincide with the centre of gravity being almost under the bar. In the average case this critical point was left rather late and for the poor case where the movement was a failure, this critical point was left even later and together with the indecisive pike, led to the failure of the movement. Hence two points emerge:

(a) a decisive pike must be made at roughly the same angle on the opposite side of the bar to the start of the movement.

(b) the straightening of the body must coincide with the subject's centre of gravity lying under the bar in the backswing part of the movement.

4 | Conserving momentum

Introduction

So far only passing mention has been made about rotational movement in previous chapters. Now it is time to study this aspect of human movement a little further.

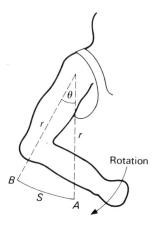

FIG. 4.1 *Rotational movement of the arm.*

If we consider the arm shown in fig. 4.1 being swung backwards, its average translational velocity (v) in metres per second will be given by the equation:

$$v = \frac{s}{t}$$

while its rotational velocity (w) will be given by:

$$w = \frac{\theta}{t}$$

where θ is measured in radians. As

$$\theta \text{ radians} = \frac{s}{r}$$

an immediate relationship between the translational component of rotational movement can be established, namely

$$v = rw.$$

This relation is true for constant as well as varying rotational velocities. In the latter case, a similar relationship exists between the translational component (a) of the rotational acceleration (α): that is:

$$a = r \cdot \alpha.$$

It follows that uniformly accelerated rotational movement is described by the equations of Chapter 1 describing uniformly accelerated linear movement, when s is replaced by θ, u by $w_{initial}$, v by w_{final}, and a by α.

As rotational velocities are sometimes given in revs/secs it should be noted that 10 revs/sec is equivalent to about 63 radians/sec.

Moments of inertia

In Chapter 2 we noted the reluctance of a body to move linearly was measured in terms of its mass, m, which is a constant parameter for any given body. We would similarly expect a body to be reluctant to turn and this is found to be the case. However, the person standing on the turntable in fig. 4.2(a) is much less reluctant to turn than the person in fig. 4.2(b) and in fig. 4.2(c). Thus the reluctance of a body to turn is not only a function of its mass but also the distribution of that mass about the axis of rotation. The reluctance of a body to turn is called its moment of inertia I. Consider the rotating sphere of fig. 4.3.

For the linear aspect of its motion we can write

$$F = m \cdot a$$

ignoring the spindle's mass. The couple, G, produced by the force F causing the rotation is given by

$$G = F \cdot r$$

which by substitution for F gives

$$G = m \cdot a \cdot r.$$

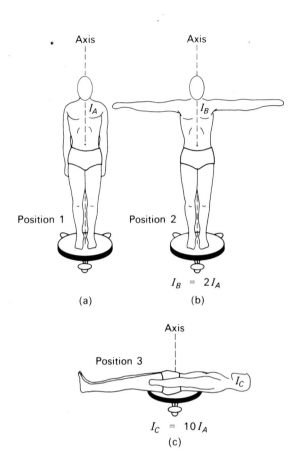

(a)

(b) $I_B = 2I_A$

(c) $I_C = 10I_A$

FIG. 4.2 *Variation in moments of inertia of the human body.*

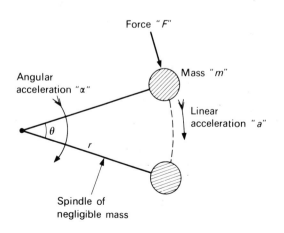

FIG. 4.3 *A rotating sphere.*

But the linear acceleration of the sphere is related to its rotational acceleration by

$$a = r.\alpha.$$

Therefore

$$G = mr^2.\alpha$$

substituting for "*a*" and this becomes

$$G = I.\alpha$$

if we identify the product mr^2 with the reluctance of the sphere to rotate, that is, with its moment of inertia I. In other words, the moment of inertia of a small sphere rotating about an external axis is given by:

$$I_{\text{sphere}} = mr^2.$$

With the apparatus shown in fig. 4.4 it is possible to check the equation $G = I\alpha$ in that, if the masses are altered in value and position such that the product "mr^2" is kept constant, the reluctance of the system to rotate should remain constant

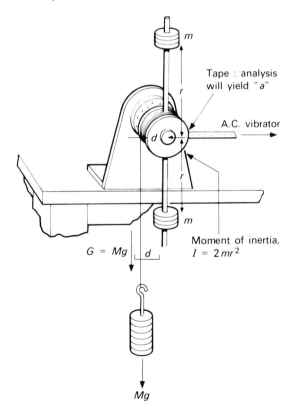

FIG. 4.4 *Apparatus for testing* $G = I.\alpha.$

so that the ratio G/α should be the same for each run. For the apparatus shown

$$G = Mg.d$$

and $\quad \alpha = \dfrac{a}{d} \quad$ Therefore $\quad \dfrac{G}{\alpha} = \dfrac{Mgd^2}{a}$

Thus the ratio Mgd^2/a should be the same for each run, or as M, g, and d are constant, the acceleration "a" measured by the ticker tape should be the same in each case. This is found to be so within the experimental error of the apparatus.

For bodies other than the simple systems of figs. 4.3 and 4.4, the moment of inertia can either be found experimentally or theoretically by summing up the product "mr^2" for each small segment of the body involved. This is a process of integration and when applied to a body like a cylinder the results of Table 7 are obtained. The axes referred to are illustrated in fig. 4.5. As the results can always be expressed in terms of the mass of the body times a dimension of the body squared, we often call this dimension the radius of gyration "k".

One result in this field that is very useful is summarised by the following equation which is applicable to any system.

$$I_A = I_G + M.h^2$$

| Moment of inertia of body about any axis. | Moment of inertia of body about a parallel axis through the body's centre of gravity. | Product of the body's mass and the square of the perpendicular distance separating the axes. |

Table 7 agrees with this equation; take for example I_{AB} and I_{CD}:

$$I_{CD} = I_{AB} + Mh_{AC}^2$$
$$= \dfrac{Mr^2}{2} + Mr^2$$

as h_{AC} is equal to r.

$$= \dfrac{3Mr^2}{2}$$

which checks with the value of I_{CD}.

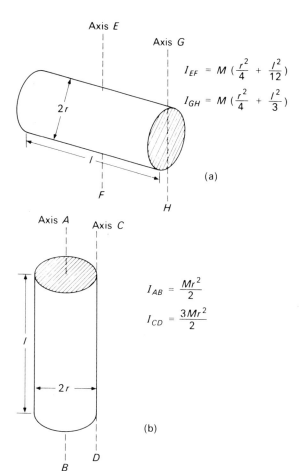

FIG. 4.5 *Moments of inertia of a cylinder.*

Table 7

Axis	Moment of inertia about given axis	(Radius of gyration)^2
AB	$\dfrac{Mr^2}{2}$	$\dfrac{r^2}{2}$
CD	$\dfrac{3Mr^2}{2}$	$\dfrac{3r^2}{2}$
EF	$M\left(\dfrac{r^2}{4} + \dfrac{l^2}{12}\right)$	$\left(\dfrac{r^2}{4} + \dfrac{l^2}{12}\right)$
GH	$M\left(\dfrac{r^2}{4} + \dfrac{l^2}{3}\right)$	$\left(\dfrac{r^2}{4} + \dfrac{l^2}{3}\right)$

Variations in moment of inertia of human body

As the human body is not a rigid body it does not have a unique moment of inertia about any given axis as, say, a wooden cylinder does. Fig. 4.2 shows three body positions about a vertical axis.

The ratio of moments of inertia between positions 1 and 2 is found to be 1:2 while the value of the ratio of positions 1 and 3 is found to be 1:10.

In general, of course, the more spread out the body is about any particular axis of rotation, the greater is its moment of inertia. It should be noted that in the air the axis of rotation will always go through the body's centre of gravity, while on the ground it will go through the point of contact. This explains why in a dive based on a toppling technique, a diver's rotation goes up 6-fold on leaving the board, for the ratio of his moment of inertia about the board to his moment of inertia about his centre of gravity is on average 6.4.[25] By the conservation of momentum principle this produces the increase in rotational speed already mentioned.

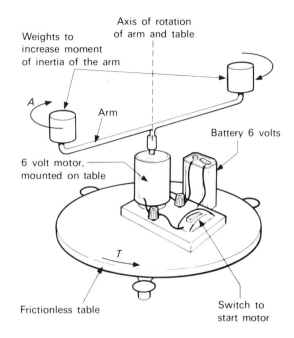

FIG. 4.6 *When the motor is started, if A represents the rotation of the arm, T will represent the rotation of the table on its frictionless bearings.*

Conservation of angular momentum

Consider the apparatus of fig. 4.6. When the motor is turned on and starts rotating the arm in a clockwise direction, the frictionless table to which it is fixed starts rotating in the opposite direction. This is but one example of the conservation of momentum principle which states that in the absence of any external restraining couple, the angular momentum of a system must remain constant.

This result is not surprising, as a similar result was given for linear motion in Chapter 2. In the rotational situation the angular momentum will be the product of the moment of inertia (I) and the angular velocity (w). For the situation shown in fig. 4.7, by the conservation principle:

$$I_1 w_1 = I_2 w_2.$$

But as I_1 is three times the value of I_2, w_2 must be three times larger than w_1. This is essentially how the pirouetting skater gains her increased rotation. Fig. 4.8 shows two other conservation effects.

It must be stressed that conservation takes place about the axis of rotation and although in many cases this lies in the vertical, medial or transverse plane, it need not necessarily do so. Fig. 4.9 shows an example where this is the case. However, as momentum is a vector quantity it can be broken down into vertical, medial and transverse components, and each of these will be conserved.

Returning to the demonstration with the turntable and rotating arm, if the turntable is rotating in an anti-clockwise direction before the motor is turned on, the action of the arm rotating in a clockwise direction would be to speed the turntable up, in order to keep the total momentum constant. Alternatively, if the turntable is rotating in a clockwise direction, turning the arm on in a clockwise direction would slow the turntable down.

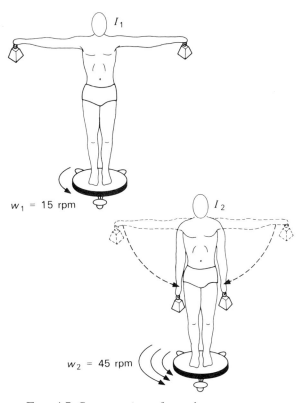

FIG. 4.7 *Conservation of angular momentum.*

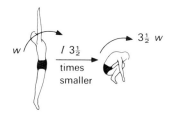

(a) Diver somersaulting

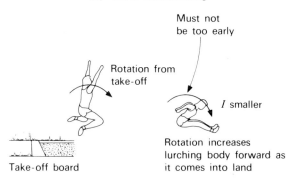

(b) Long jumping

FIG. 4.8 *Conservation of angular momentum.*

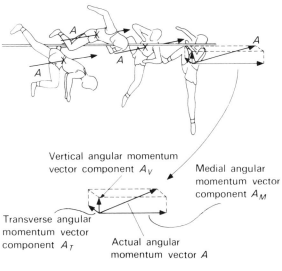

FIG. 4.9 *The straddle jump.*

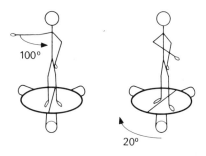

FIG. 4.10 *Controlling movements.*

Similar results can be observed with a person standing on a frictionless table.

As the subject's arm is swung in an anti-clockwise direction, angular momentum is produced in this direction. In order to keep the total angular momentum zero as it was before the arm was moved, the body swings in a clockwise direction to produce a counterbalancing angular momentum, stopping as soon as the arm does. So the net result is a stationary body that has turned through a small angle. If the subject had been rotating in a clockwise direction before the arm was swung, the movement of the arm would have speeded up the rotation momentarily. Similarly, if the subject had been rotating in an anti-clockwise direction, he would have been slowed down during the arm's movement. Actions like this, therefore, have a controlling effect on movement.

One word of warning ought to be made with respect to the conservation of angular momentum before leaving it. For a non-rigid body like the human body, if the moment of inertia of the body and its angular speed are determined while the former is changing, the conservation principle can easily be invalidated. For the human body therefore, both the moment of inertia and the angular speed must be determined over a period in which the former is reasonably constant—that is, between film frames where the relative positions of the various limbs have not changed very much.

Some human movement examples of the conservation principle

(a) *The Trampoline Twist*

As soon as the trampolinist leaves the trampoline, the arms are swung in a clockwise direction which produces an anti-clockwise rotation of the body. The arms are then moved to the side of the body, producing a small upward movement of the body. Then the twist is taken out of the shoulders which, as their moment of inertia is very much less, only brings the body back a fraction of its original rotation.

(b) *The Western Roll*

Fig. 4.11 illustrates how the conservation principle of angular moments helps the high-jumper to clear the bar using this technique.

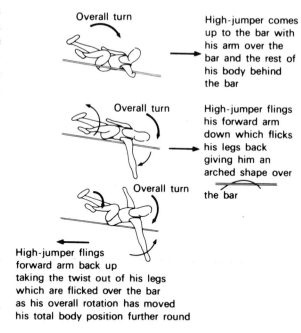

Fig. 4.11 *Western Roll high jump.*

Fig. 4.12 *The hitch kick.*

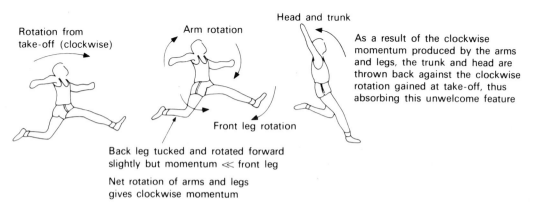

Note:- (a) If the rotation of the head and trunk < forward rotation gained at take-off, the hitch-kick slows this rotation down
     (b) If the rotation of the head and trunk is equal to the forward rotation gained at take-off the body appears to have lost its rotation and hangs in the air
     (c) If the rotation of the head and trunk is > forward rotation gained at take-off, the hitch-kick will flick the body backwards

(c) *The Tightrope Walker*

If a tightrope walker begins to fall he can correct his fall by rotating his pole or outstretched arms in the same direction as his fall. This action produces momentum in the direction of the fall which by the conservation principle produces momentum in the tightrope walker's body which opposes his fall. Providing he has not fallen too far, this will restore him to a vertical stance.

(d) *The Hitch Kick*

This activity is illustrated in fig. 4.12. As the back leg is flexed it has a smaller moment of inertia than the forward leg, so that as the latter is flung backwards, a net clockwise rotation is produced by the jumper's legs and arms. This is counter-balanced by an anti-clockwise rotation in the trunk and head. Thus, if the body has too much forward rotation the hitch kick can "take it up". If the rotation of the head and trunk equals the forward rotation at take-off, the jumper will appear to hang in the air.

(e) *The Sprinter*

In Chapter 2 we discussed the "roll" the ground thrust force produces for a sprinter. In order to reduce the effect, the sprinter rotates his arms and free leg to create momentum in the direction of the roll, which produces momentum in the rest of the body against the roll. Also by always moving one arm and leg forward when the other arm and leg are moving backwards, rotation about the vertical axis is checked.

Comparing and measuring moments of inertia

The manikin method

This method is suitable when quick comparisons are required with no great need for high accuracy. An artist's lay figure is used in place of the human body and the need for large-scale apparatus is eliminated straight away. If symmetrical comparisons about a vertical axis through the body are needed, they can be made directly. Fig. 4.13 illustrates the arrangement. The lay figure is set oscillating about its vertical axis in the first position of interest, and the period T_1 seconds of 10-20 oscillations is noted. The figure is then adjusted to assume the second position of interest, and the period of the same number of oscillations T_2 seconds is taken. It can be shown, if I_1 is the moment of inertia of the first position and I_2 that of the second position, that

$$\frac{I_2}{I_1} = \left(\frac{T_2}{T_1}\right)^2$$

The positions must be symmetrical about the axis of rotation in order that the centre of gravity of the figure lies on the axis of rotation. There is fair agreement with the ratio found using a frictionless table. The inaccuracy of the method lies in assuming that the limbs of the lay figure are equivalent to the limbs of the human body, and

FIG. 4.13 *Manikin method of comparing moments of inertia.*

not all lay figures give as good a result as the one used above. Consequently, if this method is going to be used for analysis a lay figure with segments in direct proportion to those of the human body ought to be built, and comparison studies such as those discussed in Chapter 3 carried out. However, it should be noted that there is no difficulty in getting the manikin to hold a particular position indefinitely, which is a considerable problem for some methods described.

The frictionless table method

The apparatus necessary for this method is illustrated in fig. 4.14. The arm is a strip of wood with a plate attached to each end. One of these plates is clamped to the table so that the arm extends outwards from the table along a radius. The table shown is very free running when correctly levelled and does not need a counterbalancing arm. The Panax scaler-timer is operated by the two photo-electric cells A and B which are connected to the timer via a gating unit, so that when the beams illuminating A and B are cut in turn, the timer is first started and then stopped. The Panax scaler-timer measures to a thousandth of a second, so that the time intervals involved in this experiment are measured with no significant error. The bulbs illuminating the cells are placed about 4 cm from the cells, and under this condition they operate the cells without the room being darkened.

The procedure for this method is then as follows. The subject stands on the turntable with his arms at his side. This is used as the reference position. The table is then rotated at a reasonable speed. When the subject is relaxed, and before he becomes dizzy, the Panax scaler-timer is switched to "count". Immediately after the plate at the end of the turntable's arm has started and stopped the timer, it is switched off and the time recorded is noted. The subject is then told to put his arms into the position being investigated with respect to the reference position and the timer is then switched to "count" again. The time recorded at the end of starting and stopping the timer for this second position is then noted.

With a freely turning table, frictional effects can be ignored for the number of rotations used in this particular experiment. Under these conditions angular momentum is conserved sufficiently to write:

$$I_1 w_1 = I_2 w_2$$

where I_1 and w_1 are the moment of inertia and angular velocity respectively of the first position, and I_2 and w_2 are the same quantities for the second position. If the two times recorded consecutively by the timer are t_1 and t_2, the time taken by the arm to sweep through the two photo-electric cells in the first position of the subject is t_1 seconds, while for the second position of the subject it is $(t_2 - t_1)$ seconds.

Then

$$w_1 = \frac{\theta}{t_1}$$

$$w_2 = \frac{\theta}{(t_2 - t_1)}$$

where θ is the angle subtended by the two photo-electric cells at the centre of the turntable. Substituting those values of w_1 and w_2 into the above equation gives us:

$$\frac{I_1}{I_2} = \left(\frac{t_2}{t_1} - 1\right)$$

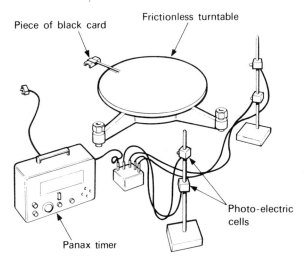

FIG. 4.14 *Frictionless table comparison.*

I_1 and I_2 represent the moment of inertia of the rotating part of the table and the subject.

Treating the table as a 2·3 kg disc and the body of the subject as a 72·6 kg cylinder in the first position, it can be shown that the ratio of the moment of inertia of the body to the table is about 14:1. It is far higher for position 2. Thus it is possible to regard I_1 and I_2 as representing the moments of inertia of the body alone without great error.

This is certainly the case for the sequence shown in fig. 4.2(a) and (b) and the results for this sequence give a ratio of 1:2. It is also possible to compare horizontal positions, for example the "tuck" position, but it is more difficult to standardise the movements and therefore results are not quite so consistent.

The segmental method

After a film has been taken of the rotational motion under study, and outlines have been made of a reasonable size from the selected frames by projecting the film through a spectrograph or slide projector fitted with a suitable gating unit, each outline is segmented into the equivalent cylinders shown in fig. 4.15. That is, each segment of the body is made into a cylinder equal in mass and length to the segment, and having a diameter equal to the average diameter of the segment found from the circumference of the segment determined at its extremities and at its centre.

The centre of gravity of each position is then found using the technique described in Chapter 3 for the hammer thrower. The combined centre of gravity of the upper arm and forearm A is found first, followed by the combined centre of gravity of the thigh and lower leg B, followed by the combined centre of gravity of the head and trunk, C, and the combined centre of gravity of A and B, D. Finally the centres of gravity D and C are linked to give the centre of gravity of the whole body, G.

The moment of inertia about G is a little more complex to find. The moment of inertia of a cylinder about an axis through G is given by:

$$I = M\left(\frac{d^2}{16} + \frac{l^2}{12}\right) + Mh^2$$

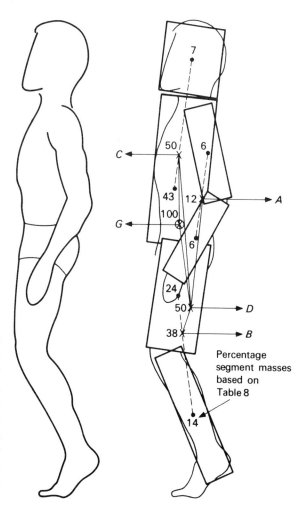

FIG. 4.15 *Centres of gravity and weights of body segments.*

where M is the mass of the cylinder, d is the diameter of the cylinder, l is the length of the cylinder and h is the distance between the centre of gravity of the cylinder and G. The first term is a constant of all the cylinders and will be referred to as I_G in future.

Knowing the I_G for all the body cylinders, allows the moment of inertia I of each cylinder about G for any given position to be found quite easily as Table 8 indicates. All that is needed are the distances h from the centre of gravity of each cylinder to the centre of gravity of the whole body G, and the masses of each cylinder, so that Mh^2 can be calculated.

Table 8

Segmental Moment of Inertia for the position shown in Figure 4.15

Segment	Distance (h)	Mass (M)	$I_G = M(\frac{d^2}{16} + \frac{l^2}{12})$	Mh^2 (kg m²)	$I = I + Mh^2$ (kg m²)
Head	0·658	4·8	0·034	2·078	2·112
Trunk	0·231	29·3	0·933	1·563	2·496
2 × Upper arm	0·320	2 × 2·0	0·034	0·410	0·444
2 × Forearm and Hand	0·046	2 × 2·0	0·058	0·008	0·066
2 × Thighs	0·267	2 × 8·2	0·254	1·169	1·432
2 × Lower leg leg and foot	0·747	2 × 4·8	0·193	5·357	5·550
Totals	Body mass	68·1	Moment of Inertia		12·100

The total moment of inertia of the body about G is obtained by adding all the separate moments of inertia together (see Table 8).

The segmentation of the body into cylinders is not without error and it can be as great as 20%. This still compares well with other methods to merit use and consideration.

Cradle method

There are basically two cradle methods that have both been used quite successfully. One is based on the compound pendulum principle that has been used by the Aerospace Medical Division in the U.S.A.[27] to determine the moments of inertia of 66 male subjects in eight body positions. The other is based on the bifilar pendulum principle and has been used by several researchers in this country to carry out moment of inertia determinations.

The two cradles discussed in this chapter are much simpler versions than those mentioned above, and the results obtained from them less accurate. However, in the context of school and college human movement projects they are quite accurate enough.

(a) *The Compound Cradle*

Fig. 4.16 illustrates a simple version of this type of cradle. The frame and cradle arms are constructed from 2·5 cm square dexion tubing. The

FIG. 4.16 *A compound cradle.*

bearings are 0·6 cm steel cones, set at the full height (2 m) and two-thirds of the height (1·33 m) of the cradle arms.

In operation, a magnetised arm attached to the bottom of one cradle arm operates a reed switch every time it passes over it. This switch in turn operates a G.P.O. relay unit which feeds a signal to a Panax timer to start it counting. Ten oscillations later another signal leaves the relay unit to stop the Panax counting. The cradle (with or without a subject strapped to it) can be switched from oscillating about the high suspension bearings to oscillating about the low suspension bearings by the lever arrangement which is operated by the hangman's noose.

For such a cradle the period of swing is given by the equation:

$$T = 2\pi \sqrt{\frac{I}{Mgh}}$$

where I is the moment of inertia of the system, M its mass, h the distance from the suspension axis to the centre of gravity of the system and g the acceleration due to gravity.

By the application of this equation and the principle of moments to the cradle by itself and with a subject strapped to it, about both axes, equations can be derived that link the distance of the subject's centre of gravity from either axis and his moment of inertia about a transverse axis through his centre of gravity to measurable parameters of the subject and the cradle.

Obviously the method has its drawbacks. First, only a limited number of positions can be studied; secondly, the movement of the subject between the high and low axis can disturb his position and cause error. He must, therefore, be firmly strapped to the cradle. In both respects the trifilar cradle is better.

(b) *The Trifilar Cradle*

Fig. 4.17 shows a simple cradle of this type.[28]

The platform, made of 5 cm solid methane foam sandwiched between two hardboard sheets, is 1·22 m wide and 2·15 m long. It is suspended by three steel wires 274 cm long which are strong enough not to appreciably extend when a subject is on the board. This keeps the moment of inertia of the cradle small which helps to increase the

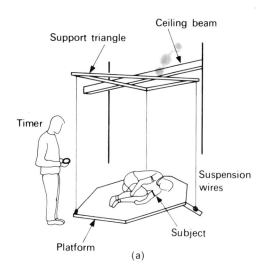

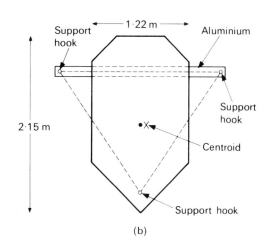

FIG. 4.17 *A trifilar cradle.*

accuracy of the method. The centre of gravity of the cradle's platform must coincide with the centroid of the supporting hooks and when the subject takes up his position on the platform, his centre of gravity must also lie under this point.

For this to occur a centre of gravity determination using a gravity board is first carried out for the required moment of inertia position and then the outline used as part of this determination is thrown on to the cradle's platform, and the latter is moved until the centre of gravity of the platform coincides with that of the projected outline.

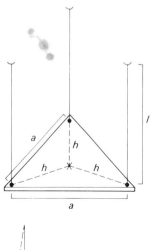

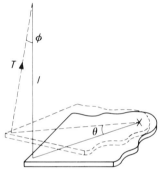

FIG. 4.18 *Diagrammatic view of trifilar cradle.*

The outline is then chalked on to the platform cradle, the platform is suspended from the three supporting wires, the subject assumes his position in the chalk outline, the cradle is set oscillating and 40 oscillations are timed. The period for the unloaded cradle is also needed together with the length of the support wires (l) and the horizontal distance separating the wires (a).

It can be shown by resolving the tension in the suspension wires vertically and horizontally, that the restoring couple G, when the cradle receives a small horizontal displacement θ, is proportional to θ, namely:

$$G = \frac{M_0 g a^2}{3l} \cdot \theta$$

where M_0 is the cradle's mass and g is the acceleration due to gravity. But

$$G = I_0 \cdot \alpha$$

where I_0 is the moment of inertia of the cradle about a vertical axis through its centre of gravity. By combining these equations we have:

$$\theta = \frac{3I_0 l}{M_0 g a^2} \cdot \alpha$$

This will be identified by readers as an equation representing harmonic motion of period

$$T_0 = 2\pi \sqrt{\frac{3I_0 l}{M_0 g a^2}}$$

so that we can write

$$I_0 = \frac{M_0 g a^2 \cdot T_0^2}{12\pi^2 l}$$

or

$$I_0 = k M_0 T_0^2$$

where k is a constant dependent only on the length and separation of the suspension wires of the cradle.

When a human being is placed on the cradle with his centre of gravity coincident with that of the cradle, the combined moment of inertia is given by

$$I_1 + I_0 = k(M_1 + M_0)T_1^2$$

where I_1 is the moment of inertia of the human being on the cradle, M_1 his mass, and T_1 the period of the cradle under these conditions. Thus:

$$I_1 = k(M_1 T_1^2 + M_0(T_1^2 - T_0^2)).$$

The largest error in this equation is caused by $M_0(T_1^2 - T_0^2)$. By making the cradle as light as possible this term is reduced in its influence. Also as the periods T_0 and T_1 are governed by "l", long suspension wires help the subject hold his position and thereby reduce damping by involuntary movement as well as lengthening the timing of the cradle's oscillations.

If a comparison of moments of inertia is required, k is eliminated, and

$$\frac{I_2}{I_1} = \frac{M_1 T_2^2 + M_0(T_2^2 - T_0^2)}{M_1 T_1^2 + M_0(T_1^2 - T_0^2)}$$

Using this technique, Philpot[28] has investigated the tucked somersault and for a male student the values he obtained for the initial position and the tucked position were 14·6 and 6·9 kg m² respectively giving a ratio of 0·47.

5 | The physics of ball games

An introduction to the centre of percussion

The concept of the centre of percussion has relevance to human movement that is sometimes overlooked because it is a more difficult concept to explain than for example the concept of the centre of gravity. Perhaps the easiest way to introduce the concept is as follows. Consider the simple system shown below in fig. 5.1.

Suppose a force F_1 is applied at A of magnitude F. This force will cause the system to rotate about its centre of gravity G in a clockwise manner as well as to translate it to the left. This can easily be verified by hitting a ruler on a desk, applying the blow at one end of the ruler. This rotational and translational movement can be more easily seen if F_1 acting at A is thought to set up two equal and opposite forces at the centre of gravity G, F_2 and F_3 each of magnitude F, as shown in fig. 5.2. This system is equivalent to that of fig. 5.1, as F_2 and F_3 cancel each other out, but it can be seen that F_1 and F_2 set up a couple of magnitude $F.x$ that causes a clockwise rotation about G, while F_3 translates the system at G to the left. Thus at point B there is a translation to the left and a rotation to the right. If the force F_1 is delivered such that the rotational movement to the right counteracts the translational movement to the left, so that there is no movement at B, A is called the centre of percussion.

The reluctance of a system to rotate is measured by its moment of inertia and the moment of inertia of a uniform block about its centre of gravity is given by the equation:

$$I = m.k^2$$

where I is the moment of inertia about G, m is the mass of the system and k is the radius of gyration of the system. For rotation, the couple G producing the rotation is equal to the moment of inertia of the system about its centre of gravity multiplied by the angular acceleration. Thus:

$$G = F.x = I.\alpha = mk^2\alpha$$

where α is the angular acceleration of the system. For translation the force producing the movement is equal to the product of the mass of the system and its translational acceleration. Thus:

$$F = m.a$$

where a is the translational acceleration of the system. Re-arranging the first equation gives the angular acceleration as:

$$\alpha = \frac{F.x}{m.k^2}$$

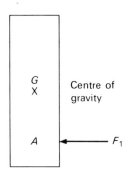

FIG. 5.1

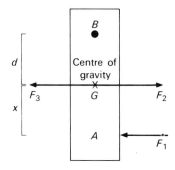

FIG. 5.2 *The centre of percussion.*

and by re-arranging $F = ma$ the translational acceleration as:

$$a = \frac{F}{m}$$

For no movement at B, the translational component of α to the right which is given by $\alpha.d$ must equal the translational acceleration a to the left. In other words:

$$\alpha.d = a$$

or

$$\left(\frac{Fx}{mk^2}\right).d = \frac{F}{m}$$

so that

$$x = \frac{k^2}{d}$$

This is the condition for no rotation at B when the system receives a blow F_1 that is at right angles to BGA. If F_1 is not perpendicular to BGA rotation will occur at B. Let us now look at some examples where this concept plays an important role.

The cricket bat

In attacking strokes in cricket, a cricketer wants to hit the ball near to the centre of percussion of the bat, so as to avoid as much jarring as possible. Treating the cricket bat as a rectangular block with a handle as shown in fig. 5.3, let us suppose the bat is gripped at B. By comparison with the system discussed earlier and represented by fig. 5.1, it can be seen that:

$$d = \frac{l}{2}$$

where l is the length of the blades. For a rectangular block the moment of inertia about its centre of gravity is given by:

$$I = m\frac{(l^2 + b^2)}{12}$$

where b is the breadth of the rectangular block and l is its length.

For a rectangular block simulating a cricket bat, l is about 50 cm and b is about 2·5 cm. Thus b^2 can be ignored with respect to l^2 and the equation reduces to:

$$I = \frac{m.l^2}{12}$$

which gives a value for k^2 of:

$$k^2 = \frac{l^2}{12}$$

Applying the condition given for no movement at B by substituting values of d and k, we have that:

$$x = \frac{l}{6}$$

This means that the centre of percussion lies two-thirds of the way down the blade of a cricket bat. In practice, because a cricket bat is not rectangular, and the grip is never at one point at the bottom of the handle but spread over an area near the base of the handle, and the moment of inertia of the handle has been ignored, the centre of percussion is lower than this and should lie at the meat of the blade. Of course, the centre of percussion of a cricket bat can be found experimentally and this will be discussed later on in this chapter.

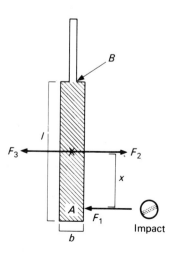

FIG. 5.3 *The cricket bat's centre of percussion.*

The tennis racket

Again for no jarring at the wrist, the position of B must coincide with the point where the player's hand grips the racket handle. As most rackets balance very close to the racket head and the length of the handle and the head are approximately the same, it is reasonable to assume that the masses of the handle and the head are equal. It is also true that the length of the handle is approximately equal to the diameter of the head so that the racket can be considered as a rod of length "l" and mass "m" attached to a hoop of diameter "l" and mass "m" as shown in fig. 5.4.

Now:

$$I_G = I_{\text{racket head}} + I_{\text{racket handle}} + 2m\left(\frac{l}{2}\right)^2$$

i.e.

$$I_G = \frac{m}{2}\left(\frac{l}{2}\right)^2 + \frac{ml^2}{12} + 2m\left(\frac{l}{2}\right)^2$$

$$= \frac{17ml^2}{24}$$

But

$$I_G = (2m)k^2$$

so that

$$k^2 = \frac{17 \times l^2}{48}$$

therefore

$$x = \frac{k^2}{d}$$

$$= \frac{17 \times l}{48}$$

as "d" is equal to "l" from fig. 5.4. Thus no jarring occurs when the racket hits the ball at a distance of about a third of the diameter below the base of the handle. In practice when measured experimentally it works out nearer the centre of the racket head as we have used a "simplified" racket in our treatment here; for example, a round head rather than an oval one and so on.

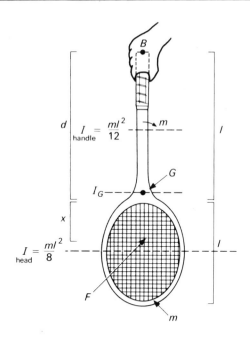

FIG. 5.4 *A tennis racket's centre of percussion.*

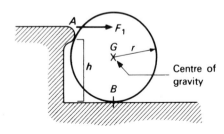

FIG. 5.5 *Centre of percussion of the billiard ball.*

For other hitting implements such as the golf club and the hockey stick the centre of percussion is moved as close to the head as possible by weighting the heads and making the shafts as light as is reasonable.

The billiard ball

For a billiard ball not to slip at B (see fig. 5.5) when it makes contact with the sides of the billiard table, A must be the centre of percussion with respect to B.

For a billiard ball:
$$I = \tfrac{2}{5}mr^2$$
where r is the radius of the ball. Therefore:
$$k^2 = \tfrac{2}{5}r^2$$
while:
$$d = r$$
so that:
$$x = \tfrac{2}{5}r$$
The height of the rim around the sides of the table indicated by h in figure 5.5 is given by:
$$h = x + d$$
$$= \tfrac{2}{5}r + r$$
$$= \frac{7r}{5}$$

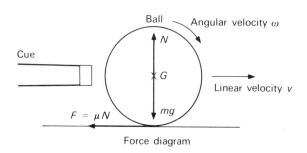

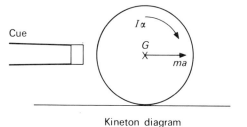

FIG. 5.6 *Dynamics of the billiard ball.*

This can easily be checked by measuring the diameter of a billiard ball with a pair of calipers and measuring the height of the rim with a ruler. A must also be the point at which the cue hits the ball if it is not to slip when hit. If the ball is hit centrally, then as soon as the cue loses contact with the ball, the only force acting on the ball is that of friction (see fig. 5.6).

Resolving forces vertically
$$N = mg$$
and thus as
$$F = \mu N$$
we can write:
$$F = \mu mg.$$

Equating the force and kineton diagrams translationally we get:
$$F = -ma$$
and substituting for F from the last but one equation
$$\mu mg = -ma$$
$$a = -\mu g$$

which represents a constant deceleration. Equating the force and kineton diagrams rotationally we get:
$$G = I.\alpha$$
or
$$Fr = \tfrac{2}{5}mr^2.\alpha$$
or
$$F = \tfrac{2}{5}mr.\alpha$$
and substituting the value of $F = \mu mg$ yet again:
$$\mu mg = \tfrac{2}{5}mr.\alpha$$
or
$$\alpha = \frac{5\mu g}{2r}$$

which represents a constant rotational acceleration. Now as long as the translational velocity is greater than the translational component of the rotational velocity, namely $r\omega$, the ball will slide. As the value of "ω" is zero when the ball starts to move, the translational velocity "v" will slow down until:
$$v = r\omega$$

and then the ball will start to roll. The ball undergoes constant deceleration so that:

$$v = u + at$$

where u is the ball's initial velocity and v is its velocity after t secs. Similarly, as the ball experiences constant rotational acceleration,

$$w = \alpha t$$

using the equivalent rotational equation and remembering its initial rotational velocity is zero. Substituting the values of "a" and "α" into these equations gives us:

$$v = u - \mu g t$$

and

$$w = \left(\frac{5\mu g}{2r}\right).t$$

Using the condition expressed above we can find the time t, and values of v and w, when the ball starts to roll. That is, substituting the values of v and w given:

$$u - \mu g t = \frac{5\mu g t}{2}$$

so that

$$\tfrac{7}{2}\mu g t = u$$

or

$$t = \frac{2u}{7\mu g}$$

and

$$v = u - \mu g . \left\{\frac{2u}{7\mu g}\right\}$$

i.e.

$$v = \frac{5u}{7}$$

while:

$$w = \left\{\frac{5\mu g}{2r}\right\}\left\{\frac{2u}{7\mu g}\right\}$$

$$= \frac{5u}{7r}$$

Thus v and w are both independent of "u" at this point while t is not, where u is the translational velocity and w the rotational velocity and t is the time the ball starts to roll.

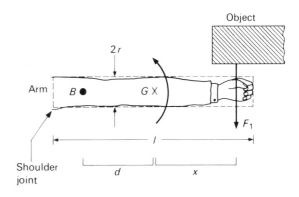

FIG. 5.7

Hitting an object with your hand

Consider the situation shown in fig. 5.7.

Treating the arm as a cylinder of length "l" and radius "r":

$$I = m\left\{\frac{l^2}{12} + \frac{r^2}{4}\right\}$$

But r is about 5 cm while l is about 60 cm, so that the $r^2/4$ term can be ignored. Thus:

$$k^2 = \frac{l^2}{12}$$

In this case x must be equal to $l/2$ which gives the following value for d when x and k^2 are substituted

$$d = \frac{l}{6}$$

This means that the point of no rotation B does not coincide with the shoulder joint and a certain amount of jarring results. A more elaborate investigation can be carried out as follows. Consider the arm shown in fig. 5.8 where the upper and lower arm are bent so that the point of no rotation coincides with the shoulder joint. For simplicity the arm is regarded in this position as two cylinders each of length "$l/2$".

The moment of inertia about the centre of gravity of the whole system is given by:

$$I = 2\left\{\frac{m}{2}\left(\frac{l}{2}\right)^2 \cdot \frac{1}{12} + \frac{m}{2}\left(\frac{l}{4}\right)^2 \cos^2\theta\right\}$$

$$= \frac{m(1 + 3\cos^2\theta)l^2}{48}$$

giving

$$k^2 = \frac{(1 + 3\cdot\cos^2\theta)l^2}{48}$$

This assumes the radius of each cylinder is very much smaller than the length.

Also:

$$d = \frac{l\cdot\cos\theta}{2} = x$$

so that $x = k^2/d$ becomes:

$$xd = k^2$$

or

$$\frac{l^2\cos^2\theta}{2} = \frac{(1 + 3\cos^2\theta)l^2}{48}$$

which gives:

$$\cos\theta = 0\cdot33$$

or:

$$\theta = 70°$$

It follows, therefore, that when playing fives, the striking arm should be bent at about 70° for there to be no jarring when the hand hits the ball.

This is obviously a simplified treatment, and in practice the angle will be less, as the impulse received and the position of the shoulder joint are not at the extremities of the arm, nor are the two sections equal in mass and length. They also have a finite radius.

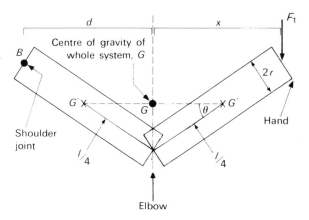

FIG. 5.8 *Shoulder jarring.*

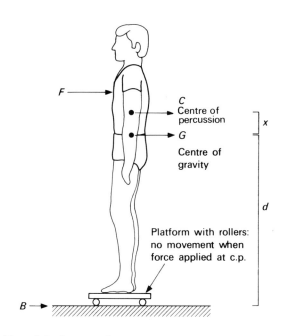

FIG. 5.9 *Centre of percussion of the human body.*

The centre of percussion of the whole body

The centre of percussion of the whole body with arms at the side can be derived remembering "k" is the radius of gyration for the whole body and "d" is the distance from the feet to the centre of gravity (see fig. 5.9). For the average young male adult,[27] "d" is 0·975 metres while his mass is 75·4 kg and his moment of inertia 11·6 kg m², so that:

$$k^2 = \frac{I}{M} = \frac{11\cdot 6}{75\cdot 4} = 0\cdot 154 \text{ m}^2$$

and therefore

$$x = \frac{k^2}{d} = \frac{0\cdot 154}{0\cdot 975} = 0\cdot 158 \text{ m}$$

This means the centre of percussion lies 0·158 metres above the centre of gravity or 1·133 metres above the feet (0·158 + 0·975).

For the sixty-six subjects for which the average values of I, M and d have just been used, the value of "x" of each individual differed by no more than 8% from the average value of 0·158 metres. It is rare to find such a consistent body parameter, and it allows us to develop another method for determining the moment of inertia of the human body, that has almost the same accuracy as the cradle methods.

FIG. 5.10 *Experimental determination of a cricket bat's centre of percussion.*

We can write the percussion equation as:

$$k^2 = x\cdot d.$$

But:

$$I = M\cdot k^2.$$

Thus:

$$I = M\cdot x\cdot d$$

substituting for "k^2". For the akimbo position of fig. 5.9 we know the value of "x" is 0·158 m for an accuracy of up to 8%, so that

$$I_{akimbo} = 0\cdot 158 \, Md$$

"x" is found to have the same constancy for other body positions providing the subjects are similar in body build to the 66 young male adults from which the original data was derived. This method certainly uses far less expensive equipment than the cradle methods, needing only scales to measure M and a c.g. board to determine d. When using a pneumatic drill it is important that the line of action of the drill passes through the centre of percussion of the whole body.

Experimental determination of centres of percussion

So far we have discussed "theoretical" centres of percussion for a number of hitting implements as well as the human body itself. This section will outline experimental methods for locating the centre of percussion, first for hitting implements such as the cricket bat and second for the human body.

Experimental determination of the centre of percussion of hitting implements

The centre of percussion can be determined practically using the apparatus illustrated in fig. 5.10. This apparatus consists of a frame to support the bat and a sliding bar, which can be drawn back and released, that hits the face of the bat. The base of the handle of the bat is held in a clamp to which two spindles are attached with dexion wheels at each end. These spindles rest on two ledges which can be raised or lowered, thus enabling the position of the bat to be altered in relation to the sliding bar. The bar is spring-loaded with two strong elastic cords. A series of experiments can then be carried out in which the bar is released from the same point to hit the face of the bat and the backward movement of the spindles along the ledges is measured. By adjusting the position of the bat in relation to the bar, an ultimate position is found where the linear movement of the spindles is least. The area struck by the bar at this ultimate position will be the centre of percussion of the bat. An alternative technique is to mount the bat in a steel plate which has strain gauges strapped to it so that the plate grips the bat at B. The point where the strain gauges give their lowest reading will be the centre of percussion of the bat under test. This will normally be a band rather than a line across the blade of the bat.

Once this position has been found, several aspects of cricket can be investigated. First, the centre of percussion can be checked against the high point of the meat of the blade. For cheaper bats, there is often quite a large discrepancy which is not constant from bat to bat of the same size and make. Secondly, as hits can be recorded by placing a sheet of tracing paper over the blade with a sheet of carbon paper sellotaped over the top of it, it is possible to investigate the use of the bat in play with respect to the centre of percussion. The centre of percussion area can be marked on the tracing paper and the carbon paper will record the nearness of hits to this area. If a cricketer is a good batsman with a reasonable drive stroke, he will tend to hit the ball much more consistently within the area of the centre of percussion than a batsman who does not have a good drive stroke. Again, if a cricketer is not an acknowledged batsman, he is likely to have a much more random display of results when his record is analysed. This technique will allow a

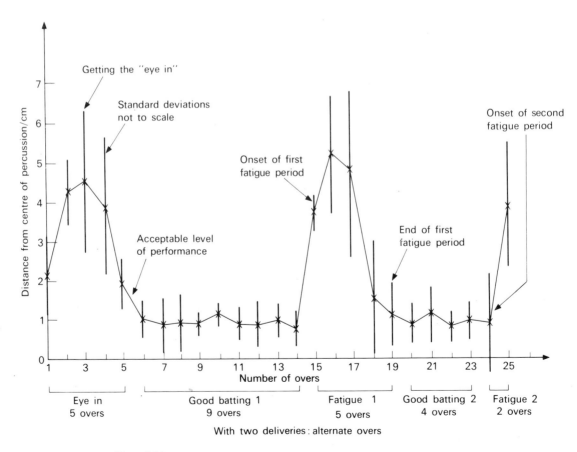

FIG. 5.11

full over to be recorded without analysis becoming too difficult. Using this technique, it is possible to investigate the following aspects of cricket:

Investigation (a) the number of overs a batsman takes to "get his eye in" so that he has accommodated his stroke to hitting the ball near to the centre of percussion of either a familiar or unfamiliar bat.

Investigation (b) the variation in the centre of percussion accuracy during a warm-up in the nets using a single stroke.

Investigation (c) the influence of bat size and weight on the centre of percussion.

The results gained in an investigation similar to (a) are shown on fig. 5.11. Note that the standard deviation (σ) is the sum of all the distances of actual strikes from the mean strike line squared. These are not drawn to scale.

Experimental determination of the centre of percussion of the whole body

The centre of percussion of the whole body where the point of no rotation coincides with the feet can be found experimentally as follows. The subject stands on a platform which itself is standing on rollers, and is asked to hold a fairly heavy bar in his hands (see fig. 5.12). He is then asked to give it a sharp push away from his body starting at hip level and then repeating the exercise raising the bar each time. When the bar is pushed away at hip level it will be below the centre of percussion and so the subject's feet will tend to move backwards. When it is pushed away at shoulder height, it will be above the centre of percussion and the subject's feet will tend to move forwards. Somewhere in between these two positions the subject will make a push that corresponds to his centre of percussion with respect to his feet, and his feet will move neither forward nor backward. In this position the platform will remain stationary. Since the subject's arms are not at his side, the centre of percussion found by this technique will differ slightly to that derived for a subject with his arms at his side.

Introducing the coefficient of restitution

When a cricketer hits a ball with his bat he changes the direction of its motion. Thus, on being hit, the ball deforms itself until it brings itself momentarily to rest, storing up energy as it does so and then uses this energy to spring away from the bat, regaining its original shape in the process. Not all the stored energy is converted back to kinetic energy; some is converted to heat, so that the ball's temperature will rise slightly during impact. Although this is hardly noticeable with a cricket ball, the effect is very noticeable with squash balls. Of course, in some cases, like that of a tennis player hitting a ball, both the ball and the hitting implement undergo deformation.

It is found that, for a particular hitting implement and ball, the velocity of approach before impact and the velocity of separation after impact is governed by the equation

$$e = \frac{\text{velocity of separation}}{\text{velocity of approach}}$$

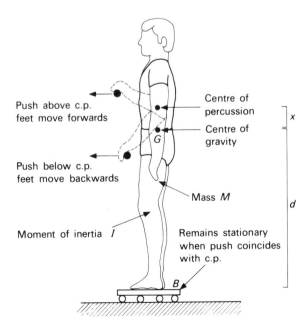

FIG. 5.12 *Experimental location of the centre of percussion.*

where 'e' is a constant for a particular pair of surfaces, and is called the coefficient of restitution.

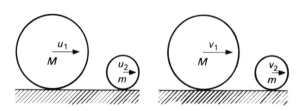

FIG. 5.13 *Coefficient of restitution.*

Thus for the situation shown in fig. 5.13 where $u_1 > u_2$ and $v_2 > v_1$ for impact and separation to take place:

$$e = \frac{(v_2 - v_1)}{(u_1 - u_2)}$$

Official standards

Consider a ball dropped from a height h_1 on to a solid floor, which after impact rises to a height h_2. Neglecting air resistance, the kinetic energy just before impact on the downward fall will be equal to the loss in potential energy, namely

$$\tfrac{1}{2}mu_1^2 = mgh_1$$

so that

$$u_1 = \sqrt{2gh_1}$$

Similarly after impact the kinetic energy left will all be transformed back into potential energy so that

$$\tfrac{1}{2}mv_1^2 = mgh_2$$

so that

$$v_1 = \sqrt{2gh_2}$$

Now in this example, u_2 will be equal to v_2 and will be zero, and v_1 will be in the reverse direction to that shown in fig. 5.13.

Thus:
$$e = \frac{-(-v_1)}{u_1}$$
$$= \frac{v_1}{u_1} \quad \text{or} \quad e = \sqrt{\frac{h_2}{h_1}}$$

To conform to official standards a tennis ball must rise to a height of 106 to 116 cm when dropped from a height of 2 metres. Thus "e" must range from:

$$e = \sqrt{\frac{106}{200}} \text{ to } \sqrt{\frac{116}{200}}$$

i.e.

$$e = 0.73 \text{ to } 0.76$$

Table 9 gives the values of "e" for several different balls used in various games. The balls were dropped on concrete in each case.

Table 9

Ball	Coefficient of restitution when dropped on a hard floor
Tennis	From 0.73 to 0.76
Golf	Not greater than 0.70
Cricket	About 0.3
Basket ball	From 0.74 to 0.78
Table tennis	About 0.8
Hockey	About 0.6

The physics of ball games

A full discussion of this topic is beyond the scope of this book and our digression into it has been occasioned by the bearing it has on human movement. It is also unnecessary to proceed any further, even if space allowed, for C. B. Daish[29] has written an excellent book that covers this area in a very full and readable way. Any reader who wishes to pursue this topic further is recommended to buy this book.

However, as work of the type described in this book is frequently dismissed as being academic and of no practical purpose, we should like to quote an experiment carried out by Daish that shows the contrary. Professional opinion in golf has often recommended heavier club-heads than normal, to increase the speed of the club-head at impact in a drive, especially for stronger players.

By taking several subjects with different handicaps, Daish established a power law of the form

$$Mu_1^n = C$$

for each subject, where M is the mass of the club-head, u_1 the velocity of the golf club head just prior to impact, and n and C are constants. The value of C was found to vary according to the player's build and handicap, but n was found to be approximately constant for all the players with a value of about 5.

During the drive, momentum must be conserved, so referring to fig. 5.13 and remembering u_2 will be zero for a golf drive

$$Mu_i = Mv_1 + mv_2$$

But by the coefficient of restitution equation

$$eu_1 = v_2 - v_1$$

Eliminating v_1 from these two equations gives

$$v_2 = \frac{M(1+e)u_1}{(M+m)}$$

$$Mu_1^n = C.$$

and by eliminating u_1 using Daish's equation

$$v_2 = \frac{M(1+e)}{(M+m)} \cdot \left(\frac{C}{M}\right)^{1/n}$$

or

$$v_2 = \frac{KM^{(1-\frac{1}{n})}(1+e)}{(M+m)}$$

where $C^{1/n}$ is equal to K and is, of course, a constant. If the initial golf ball velocity after impact v_2 is plotted against the mass of the club-head M, the resulting curve gives a maximum value of v_2 when

$$M = m(n-1).$$

This can be derived algebraically by equating the differential coefficient dv_2/dM to zero. Readers familiar with calculus techniques will be able to verify that

$$\frac{dv_2}{dM} = \frac{K(1+e)}{(M+m)^2}\left[(M+m)\left(1-\frac{1}{n}\right)(M)^{-1/n}\right.$$
$$\left. - (M)^{(1-1/n)}\right]$$

Equating this differential coefficient to zero means that the last bracketed term must be zero, so that we have

$$(M+m)\left(1-\frac{1}{n}\right)(M)^{-1/n} - (M)^{(1-1/n)} = 0$$

i.e.

$$(M+m)(n-1) = nM$$

multiplying through by n and dividing by $M^{-1/n}$. Thus

$$M = m(n-1)$$

as stated previously. The value of m for a golf ball is 0·05 kg approximately, so that with Daish's value of 5 for n, we have

$$M = 0·05 \times 4$$

thus

$$M = 0·2 \text{ kg}$$

which agrees well with the normal club-head mass, and argues firmly against the view that heavier club-heads give better results for more powerful players.

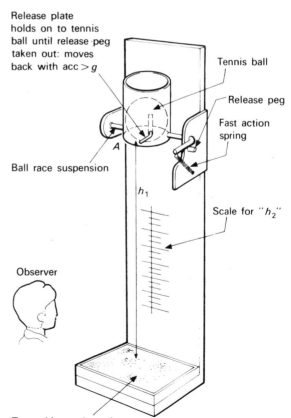

FIG. 5.14 *Measuring the coefficient of restitution.*

The measurement of the coefficient of restitution

Fig. 5.14 shows a very simple device for measuring the coefficient of restitution for tennis and similar balls. It allows them to be tested on either a standard bed (such as concrete) or any surface that is pertinent to the game in which the ball is involved. The release mechanism that lets the ball drop moves away with an acceleration greater than that due to gravity, so the ball fall commences at A. An observer positions himself such that he can read off the height the ball reaches after bouncing on the scale provided (h_2). The distance dropped before bouncing is, of course, fixed (h_1) so that "e" can be calculated directly from the equation. Of course, most balls undergo a change in the coefficient of restitution with usage. However, with the tennis ball this change is rather different to that experienced with other balls. This is because the conventional tennis ball is pressurised with gas and this gas leaks out with usage and with time, so that the bounce of the tennis ball is reduced. In addition, as the tennis ball cover wears, the deadening effect of the cover is reduced and therefore the bounce of the tennis ball in this respect increases! As stated earlier, new balls must bounce between 106 and 116 cm when dropped from a height of 2 metres. As gas leaks out (but the cover does not wear) the rebound reduces to about 96 cm. Alternatively, if the cover is worn down rapidly with the cover "as new" the rebound goes up to about 130 cm. In terms of the coefficient of restitution this means:

e (96 cm rebound from 200 cm) = 0·69

e (130 cm rebound from 200 cm) = 0·80.

Suggestions for projects and investigations

1. Using the ticker-tape apparatus described in Chapter 1, compare (a) the efficiency of the bunch and crouch spring starts and (b) the horizontal force patterns developed by a swimmer using different styles.

2. By taking a cine film of a back somersault and using the trifilar cradle technique of Chapter 4, determine the moment of inertia of six frames. Then by using frames immediately before and after the frames used to determine the moments of inertia estimate the angular velocity of each position and check that the conservation of momentum principle is valid.

3. Using one of the stroboscopic techniques outlined in Chapter 1, record the strokes of (a) a batsman hitting an *airflow* ball (b) a golfer hitting a *ping-pong* ball. By velocity analyses determine any "weakness" of style from the delay in achieving maximum velocity and hitting the ball. Does the weight of the bat or golf club head have any influence on the "weakness"? You can do this for a golf club by cutting off the head and tying a screw thread on to this end of the shaft. Ten discs of 30 g mass, centrally tapped so that they can be screwed on to the end of the golf club shaft, will give you a useful "weighted head" range.

4. By setting up the apparatus in fig. 6.1 investigate the drag resistance acting on the lay figure over the period of one stroke and for different stroke styles, using a wind generator. Produce a hierarchy of styles in terms of streamline efficiency, and comment why a high-order style may not be a particularly fast one. (Does a good streamline style necessarily allow a swimmer to develop a good forward thrust style?)

5. Design and build either a mechanical force platform or one using strain gauges that will allow you to measure the *maximum* force developed during a standing jump. By using a rotating drum and mechanically/electrically operated pen modify your original platform so that it can give

FIG. 6.1 *Drag resistance apparatus.*

a continuous trace. This will then allow you to carry out investigations like those of Howard Payne mentioned in Chapter 2.

6. Using the centre of gravity board technique of Chapter 3 produce a c.g. trace from a side elevation cine film of any gymnastic or trampoline movement. Use the trace to:

(a) verify the path taken by the centre of gravity is parabolic for the movement that takes place through the air.

(b) estimate the angular velocity at various stages and check that as the movement of inertia of the performer decreases, his or her angular velocity increases, or vice versa.

(c) for any part of the movement on the ground, determine the increase in the vertical and the horizontal velocity at various points so that the acceleration and thrust exerted can be estimated at these points in magnitude and direction.

7. By filming a good, bad and indifferent gymnastic movement on the ropes or high bar and using the resulting centre of gravity traces, determine the likely weak points in the movement.

8. Using a frictionless table compare the moment of inertia of a tucked position with an extended position of a somersault drive. What average value do you get for 20 subjects? Is this what you would expect?

9. By dropping various balls (e.g. tennis ball, basket ball, ping-pong ball) from a given height and noting the height to which they bounce, measure their coefficient of restitution. Note how wear and tear and/or loss of pressure affects this characteristic. Tennis balls for example can be worn down artificially by rubbing them down with sandpaper, and depressurised by piercing them with a needle. As an extension of this work a basket-ball floor can be analysed for "weak spots" by detecting where the drop in the coefficient of restitution falls below the average for the floor.

10. By using the technique described in Chapter 5 determine whether the centre of percussion does lie at the "meat" of the blade (for cheap, medium priced and expensive bats). Also by covering the blade of a bat with carbon paper placed face down on some white paper, see how the point of contact with the ball varies over 30 overs in the net.

(a) with one standard delivery
(b) with three standard deliveries
(c) changing bats every five overs.

You will need to change the white paper every over and the central horizontal axis of the paper must coincide with the bat's centre of percussion for analysis purposes. What conclusions can you draw from your results about the usefulness of a warm-up in the nets before play, and of using your own bat?

11. The retardation by air drag on a ball travelling below its critical velocity is proportional to ($diameter^2$). The inverse of this ratio can therefore be taken as an estimate of the ball's flight properties. This parameter can then be used to set up a league table for cricket, hockey, baseball, golf, squash, tennis and table tennis balls. Does such a table compare well with the experimental results found from throwing these balls from a catapult device at a standard angle and with a uniform force?

Anyone wishing to develop this sort of project work further should consult *Project Technology Briefs* published by Heinemann for the Schools Council, back numbers of the *School Technology Journal* published by the National Centre for School Technology, Trent Polytechnic, Burton Street, Nottingham, or *Man and Machines* by the present author, published by Pergamon in their Projects in Physics series.

References

1. Hoffman, H., 'Biomechanical investigation of the long jump and triple jump', *Boletin 1/1966 Cuba*, Catedra de Biomecanica Exuelo. Superior de Educacion Fisica, Habana, Cuba.
2. Brodie, A. D. (March 1972), 'An analytical technique for weight lifting', *British Journal of Physical Education*, Vol. 3, No. 2.
3. Daish, C. B. (September 1965), 'Effectiveness of the golf club head mass on the speed of the golf ball', *Institute of Physics Bulletin*, Vol. 16, No. 9.
4. Henry, F. M. (1952), 'Force-time characteristics of the sprint start', *Research Quarterly*, No. 23.
5. Whitney, R. J. (1958), 'The strength of the lifting action in man', *Ergonomics* 1.
6. Williams, D. M., 'A force analysis platform for work study research', unpublished MSc Thesis of Birmingham University.
7. Payne, A. H., (1966), 'The use of force platforms in the study of physical activity', *Alta 1: The University of Birmingham Review*. (See particularly page 25.)
8. Cunningham, D. M. & Brown, G. W. (1952), 'Two devices for measuring the forces acting on the human body during walking', *Proc. Soc. Exp. Stress Analysis*, No. 9.
9. Payne, A. H. & Blader, F. B., 'Design development and use of force platforms', *First Annual Report*, University of Birmingham Biomechanics Unit.
10. Payne, A. H. & Blader, F. B. (April 1970), 'A preliminary investigation into the mechanics of the sprint start', The Research Edition of the *Bulletin of Physical Education*, Vol. 8, No. 2. (In particular page 23.)
11. Lloyd Instruments Ltd, Beaumont Road, Banbury, Oxon.
12. Page, R. L. (December 1975), 'Role of Physical Buoyancy in Learning to Swim', *Journal of Human Movement Studies*, Vol. 1, pp. 190–198.
13. Von Dobeln, W. (1956), 'Human standard and maximum metabolic rate in relation to fat-free body mass', *Acta Physiologica Scandinavica*, Vol. 37, Supplement 136.
14. Akers, R. & Buskirk, E. R. (May 1969), 'An underwater weighing system utilizing "force cube" transducers', *Journal of Applied Physiology*, Vol. 26, No. 5.
15. Page, R. L. (1974), 'The position and dependence on weight and height of the centre of gravity of the young adult male', *Ergonomics*, Vol. 17, No. 5, pp. 603–612.
16. Tricker, R. A. R. & B. J. K., *The Science of Movement*, Chapter 8 and Appendix 6, Mills & Boon.
17. Web Board—C.I.B.A. (A.R.L.) Ltd, Duxford, Cambridge.
18. Dempster, W. T. (1955), 'Space requirements for the seated operator', *W.A.D.C. Report*, University of Michigan.
19. Williams, M. & Lissner, H. (1962), *Biomechanics of Human Movement*, Saunders, London.
20. Dreyfuss, R. (1955), *The Measure of Man*, 2nd ed., Whitney Library of Design and *Designing for People*, Simon & Schuster.
21. Hertzburg, H. T. (1950), 'Anthropometry of flying personnel', *W.A.D.C. Report*, University of Michigan.
22. Whitney, Dr W. J., Human Biomechanics Laboratory, National Institute for Medical Research, Hampstead Laboratories, Holly Hill, London NW3.
23. Lane, K., Lecturer in Physical Education, Physical Education Department, Edinburgh University, Pleasance, Edinburgh.
24. Harris, R. N. (1952), Body Mechanics Department, Springfield College, Massachusetts, U.S.A.
25. Page, R. L. (1973), 'The mechanics of toppling techniques in diving', *Research Quarterly*, Vol. 45, No. 2, p. 185–192.
26. Smith, J., former student of St Paul's College, Cheltenham.
27. Santchi, W. R., Dubois, J. & Omoto, C., 'Moments of inertia and centers of gravity of the living human male', *Report AMRL TDR-63-36*, Wright-Patterson Airforce Base, Ohio.
28. Philpot, A. (1974), 'Mechanics of the human body during a somersault', *Physics Education*, Vol. 9, pp. 338–342.
29. Daish, C. B. (1972), *Physics of Ball Games*, The English Universities Press Ltd.

Index

Acceleration
　constant　2
　definition of　2
　due to gravity　3
　equations for constant　3
　force and acceleration　15
　linear and angular　45
Angle of lean　15

Ball bounce
　experimental determinations　68
　official standards　66
Ball games　66
Basket ball　15
Billiards
　centre of percussion　59–60
　equations of motion　60–61
Body segmentation　38–39
Buoyancy
　definition　29
　determination of　29–30

Centre of gravity
　combination of　35, 53
　human movement and　3, 10, 13, 16, 21, 31
　movement through the air　31, 43, 71
　position of　31, 39
Centre of gravity determinations
　cradle techniques　55
　gravity board methods　35–38, 71
　manikin methods　38–43
　plumbline technique　42
　principle of moments technique　42
Centre of gravity traces　37, 43, 72
Centre of percussion
　definition of　57
　experimental determination of　63–65, 72
　introduction to　57
　of the arm　61–62
　of a billiard ball　59–60
　of the human body　62–63, 65
　of a cricket bat　58, 63
　of a tennis racket　59
Cine analysis　4, 21, 37, 42, 43, 53, 56
Circular motion
　equations of　18
　examples of　18–20
Coefficient of fraction
　equation for　14
　values of　14
Coefficient of restitution
　definition of　65
　determination of　68, 72
　equations for　66, 67
　introduction to　65
　values of　66
Conservation of momentum
　angular　48–50, 52, 71
　as a vector quantity　48, 49
　examples of　49, 50–51
　experimental verification　48, 71, 72
　in controlling human movement　49
　linear　19, 67
Contact friction　14
Couples　16, 34, 42, 45, 46, 47, 56, 57
Cricket　16, 58, 63–65, 71, 72

Deformation　13
Density of the human body　28
Diving　3, 17
Drag friction　14, 71, 72
Dynamometers　22

Electromagnetic induction　12
Equilibrium
　dynamic　35
　static　33–35
　stable　34
　unstable　35

Football kick　13, 15
Force
　and acceleration　15
　centrifugal　19
　centripetal　18
　direct measurement of　22–28, 71
　effects of　13
　frictional　14
　indirect measurement of　20–22, 72
　gravitational　13
　normal contact　13
　rotational effects of　16
　translation effects of　16
　weight　13
Force platforms
　cantilever strain gauge　24–26
　introduction to　22
　mechanical　23–24, 71
　rigid strain gauge　26–27, 71
　rigid transducer　30
　Stanmore Sandals　27–28
　underwater
　　cantilever strain gauge　29–30
　　mechanical platform　30
　　spring balance suspension　29
　　transducer platform　30
Force time graphs　16, 25, 26
Friction
　contact　14
　drag　14, 71, 72
　measurement of　14
　coefficient of　14
Frictionless table　48, 49, 52, 72

Golf drive　11, 12, 66–67, 71
Graphs　1, 3, 5, 16, 25, 26, 43, 64
Gymnastics
　giant circle　21–22
　upstart　43

Hammer throwing　18, 34
Hand stand　33
Head stand　33
Hitting
　with a bat　58
　with a cue　60
　with the hand　61
　with a racket　59
Horse racing　12

Impulse　15

Jumping
　crouch　25
　high　13, 32, 50
　long　32, 33, 49, 50, 51

Lifting loads　34

Measuring moments of inertia
　compound cradle　54
　frictionless table　52
　manikin method　51
　segment method　53
　trifilar cradle　55, 71
Methane foam board　55
Moments of inertia
　and angular acceleration　46, 57, 60
　definition of　46
　experimental confirmation　46
　introduction to　45
　of a cylinder　47, 53, 54
　of a rectangle　58
　of a sphere　45
　parallel axis theorem　47
　radius of gyration　47, 57–63
　relationship to applied couple　46, 57, 60
　values of human M of I　54
　variation of human M of I　46, 48, 53
Moment of a force　16
Momentum
　angular conservation　48, 51, 71, 72
　linear conservation　16, 67
Movement in a plane　3
Movement on the ground　33
Movement through the air　31, 43, 71
Muybridge　12

Newton's laws
　second　6, 15, 18, 20, 45, 57
　third　16, 19

Oscillations　35, 51, 55, 56

Panax timer　12, 52
Parabolic path　32, 33, 43, 71
Pendulum　35, 51, 54–56
Photo-electric cells　11–12, 52
Piking　44
Polaroid camera　10, 71
Principle of moments　35, 36, 37, 40, 71
Projects
　cricket　71, 72
　golf　71, 72
　gymnastics　71, 72
　jumping　71
　somersaulting　71, 72
　sprinting　71
　swimming　71
　tennis　72
　trampolining　71

Range　32, 33
References　69–70
Relay unit　55
Rotational movement　3, 13, 45
Running　17, 19–20

Shot-put　8, 32, 33, 35
Skating　48
Somersaulting　10, 49
Sprinting　4–6, 8, 17, 23–24, 26–27, 35, 51

INDEX

Stanmore Sandals   27
Stroboscopic analysis
   force   21
   velocity   10–11, 71
Swimming   15, 71

Tennis   11, 59, 66, 68
Ticker-tape equipment
   force analysis   20, 71
   velocity analysis   4–6, 71
Tightrope walking   33, 51

Toppling angle   34
Trampolining   50, 71
Translational movement   3–9, 13

Velocity
   analysis   5, 8, 9, 11, 12
   average   1
   constant   1
   definition   1
   instantaneous   1
   linear and rotational   45, 71

   uniformly increasing   2
Velocity-time graphs   1, 5, 71

Underwater weighing devices   28–30
Ultra-violet recorder   12, 25

Weight   13
Web board   36

Xenon lamp   10